QUAKER CONTENT

SACRED MOBILITIES

Current Developments in the Geographies of Leisure and Tourism

Series Editors:

Jan Mosedale, University of Applied Sciences HTW Chur, Switzerland and **Caroline Scarles**, University of Surrey, UK in association with the Geographies of Leisure and Tourism Research Group of the Royal Geographical Society (with the Institute of British Geographers).

Tourism and leisure exist within an inherently dynamic, fluid and complex world and are therefore inherently interdisciplinary. Recognising the role of tourism and leisure in advancing debates within the social sciences, this book series is open to contributions from cognate social science disciplines that inform geographical thought about tourism and leisure. Produced in association with the Geographies of Leisure and Tourism Research Group of the Royal Geographical Society (with the Institute of British Geographers), this series highlights and promotes cutting-edge developments and research in this field. Contributions are of a high international standard and provide theoretically-informed empirical content to facilitate the development of new research agendas in the field of tourism and leisure research. In general, the series seeks to promote academic contributions that advance contemporary debates that challenge and stimulate further discussion and research both within the fields of tourism and leisure and the wider realms of the social sciences.

Other titles in the series

Travel and Transformation
Edited by Garth Lean, Russell Staiff and Emma Waterton

Moral Encounters in Tourism
Edited by Mary Mostafanezhad and Kevin Hannam

Travel and Imagination
Edited by Garth Lean, Russell Staiff and Emma Waterton

Lifestyle Mobilities
Intersections of Travel, Leisure and Migration
Edited by Tara Duncan, Scott A. Cohen and Maria Thulemark

Sacred Mobilities
Journeys of Belief and Belonging

Edited by

AVRIL MADDRELL,
University of the West of England, UK

ALAN TERRY
University of the West of England, UK

TIM GALE
University of the West of England, UK

ASHGATE

Published by
Ashgate Publishing Limited
Wey Court East
Union Road
Farnham
Surrey, GU9 7PT
England

Ashgate Publishing Company
110 Cherry Street
Suite 3-1
Burlington, VT 05401-3818
USA

www.ashgate.com

British Library Cataloguing in Publication Data
A catalogue record for this book is available from the British Library

The Library of Congress has cataloged the printed edition as follows:
Sacred mobilities : journeys of belief and belonging / edited by Avril Maddrell, Alan Terry and Tim Gale.
    pages cm. – (Current developments in the geographies of leisure and tourism)
    Includes bibliographical references and index.
    ISBN 978-1-4724-2007-7 (hardback) – ISBN 978-1-4724-2008-4 (ebook)
– ISBN 978-1-4724-2009-1 (epub) 1. Travel–Religious aspects. 2. Tourism–Religious aspects. 3. Spiritual life. I. Maddrell, Avril, 1964– editor of compilation. II. Terry, Alan, 1954– editor of compilation. III. Gale, Tim, editor of compilation.
    BL628.8.S24 2015
    203'.5–dc23

2014036125

203·5

ISBN   9781472420077 (hbk)
ISBN   9781472420084 (ebk – PDF)
ISBN   9781472420091 (ebk – ePUB)

MIX
Paper from
responsible sources
FSC
www.fsc.org   FSC® C013985

Printed in the United Kingdom by Henry Ling Limited, at the Dorset Press, Dorchester, DT1 1HD

Contents

List of Figures

List of Contributors

Simon Arlidge is Programme Manager for the BA in Geography at the University of the West of England. Simon's teaching interests centre on managing rural landscapes for conservation and people, Australian rural studies and, more recently, renewable energies within societies. His latest research is on urban decline in outback Australia.

Katy Beinart trained as an architect at the Bartlett, UCL and Oxford Brookes, going on to develop an artistic practice that engages with the public realm. Her practice includes installations, performance and participatory works, which explores links between material culture, migration and memory, and place. She is currently working on a practice-based PhD at the Bartlett, University College London, and is a Senior Lecturer in Architecture at the University of Brighton. Her current interests are the politics and poetics of regeneration and place-making in the city, with a particular focus on migration, heritage and memorialisation.

Belén Mª Castro Fernández holds a degree in art history (1997), a Masters in pathology and architectural restoration (1998), a PhD in art history (2007), and a degree in journalism (2011). She is an associate professor at the Cesuga School of Architecture, University College Dublin (A Coruña, Spain). Her cultural heritage research and writing focuses on conservation, tourism and urbanism. She is especially interested in the Way of St. James and in the monumental twentieth-century restorations of it carried out in Spain and Portugal. Recent publications include journal articles and chapters on heritage, urban and cultural tourism, pilgrimage, benchmarking of tourism products, and the process of heritage-making in geographical space.

David Crouch is Professor of Cultural Geography in the Humanities Department at Derby, a senior member of its *Identity, Conflict and Representation* Research Centre, and an associate in cultural geography, Nottingham University. His cultural geographic research and writing includes matters of leisure, tourism, landscape and nature. His most recent publications include *Flirting with Space: Journeys and Creativity* (Ashgate, 2010), and several chapters on heritage, affect and feeling, cultural tourism, landscape studies, and artistic expressions of space and ecology. His edited books include *Leisure/Tourism Practices* (Routledge, 1999), *Visual Culture and Tourism* (Berg, 2005) and *The Media and the Tourist Imagination* (Routledge, 2005).

Tim Edensor teaches cultural geography at Manchester Metropolitan University. He is the author of *Tourists at the Taj* (Routledge, 1998), *National Identity, Popular Culture and Everyday Life* (Berg, 2002) and *Industrial Ruins: Space, Aesthetics and Materiality* (Berg, 2005), and editor of *Geographies of Rhythm* (Ashgate, 2010). He is also editor of the journal *Tourist Studies*. Tim has written extensively on national identity, tourism, industrial ruins, walking, driving, football cultures and urban materiality, and is currently investigating landscapes of illumination and darkness.

Tim Gale is Associate Head of Department and Tourism and Events Programme Leader in the Bristol Business School, at the University of the West of England. His research interests include the decline and restructuring of cold-water resorts in Britain and Northern Europe, new tourism spaces, places and experiences (for example, cyber- and eco-tourism, urban beaches), and tourism and events education. He has written and collaborated on numerous works in connection with the above (from positioning papers and literature reviews to multi-method empirical studies incorporating questionnaire surveys and interviews, document analysis and visual methods), and is particularly interested in the potential for knowledge creation offered by the 'new mobilities paradigm' and critical realism as a philosophy of/for the social sciences.

Nick Hopkins is a Senior Lecturer at the University of Dundee. As a social psychologist he is interested in how our sense of who we are is bound up with our membership of social groups. He has a longstanding interest in national and religious social identifications and how these are socially constructed so as to organise group members' behaviour. He is also interested in the social psychology of intergroup relations – especially as it concerns the relationship between majority and minority groups – and he has addressed this through studying the experiences of British Muslims. The research on pilgrims at the Prayag Magh Mela reported in this book arose out of a longstanding collaboration with the *Centre of Behavioural and Cognitive Sciences*, Allahabad University. This project sought to explore how social psychological theories of group processes can help illuminate diverse aspects of the experience of participating in mass gatherings.

Rob Irving is an artist. During the 1990s and 2000s he led what the science journal *Nature* has described as 'a growing underground art movement combining mathematics, technology, stalks and whimsy', commonly known as the crop circles phenomenon. Irving recently completed his doctoral studies at the *PLaCE* Research Centre at the University of the West of England with a thesis titled 'Playing Puck: a study of performative action in the shaping of a "legend landscape"'. His research practice draws upon an interdisciplinary framework consisting of art history, anthropology, performance theory and folklore studies, exploring the dynamics of how perceptions and experiences of 'paranormal' phenomena are influenced by place, and vice-versa. His area of speciality is ostension, or legend-telling by action, and the use of anonymity/invisibility as part of a creative process that ensures the continuity of myth as it is fed back into society as contemporary folklore.

Sammyh Khan holds a PhD in psychology from the *Centre of Applied Cross Cultural Research* at Victoria University of Wellington (New Zealand). He is currently a lecturer in social psychology at the University of Keele. Before joining Keele University, he held postdoctoral fellowships at the University of Dundee, the University of St Andrews, and the University of Exeter. His research has two key areas of focus: (a) group processes and intergroup relations, with a particular focus on religious and political mobilisation in the Indian Subcontinent; and (b) the application social identity and self-categorisation processes to understanding and promoting health behaviours and outcomes. His research employs mixed-methods, combining qualitative methodologies with survey and experimental methodologies.

Avril Maddrell is Associate Professor in Social and Cultural Geography at the University of the West of England, Bristol. Her current research interests include: spaces, landscapes and practices of pilgrimage and sacred mobilities; mourning and remembrance; geographical thought and practice; and gender. Key publications include *Christian Pilgrimage, Landscape and Heritage: Journeys to the Sacred* (Routledge, 2015, co-authored with Veronica della Dora, Alesandro Scafi and Heather Walton); *Deathscapes. Spaces for Death, Dying, Mourning and Remembrance* (Ashgate, 2010, co-edited with James Sidaway); *Memory, Mourning, Landscape* (Rodopi, 2010, co-edited with Elizabeth Anderson, Kate McLoughlin and Alana Vincent); and *Complex Locations. Women's Geographical Work in the UK 1850–1970* (RGS-IBG/Wiley-Blackwell, 2009). She is also an Editor of the journal *Gender, Place and Culture.*

Kavita Pandey obtained her PhD from the Department of Psychology, University of Allahabad. Her PhD research work focused on understanding the dynamics of retributive justice. After completion of her PhD, she worked at the *Centre of Behavioural and Cognitive Sciences*, Allahabad University, as a post-doctoral research fellow. During this time, she researched the experience of pilgrims (*Kalpwasis*) attending the Prayag Magh Mela at Allahabad. This involved interviewing *Kalpwasis* and undertaking ethnographic observations of these pilgrims before, during and after the pilgrimage event. Currently, she is working on understanding how religious identity impacts understandings of justice and its policy implications with regard to health care systems. Her research interests include social justice and social psychological dimensions of health and well-being.

Pamela Richardson leads a full and interesting academic life as an Honorary Fellow of the University of Exeter, where she was awarded a PhD in 2007 for her work on West Country Quakers, 1860–1920. She writes and speaks on Quaker subjects, published her first book, *Two Quaker Ladies: A Family Story* in 2009 and contributed a chapter to *Narrations of Travel and Tourism* (Ashgate, 2012). Her current work explores the Quaker role in World War One and a book is in process for publication later this year. She is also active in Medical History research, working on two twentieth century medical research projects, a book chapter on stress in the post-World War Two period and a refereed journal article on the development of specialisation in Orthopaedic Nursing.

Suha Shakkour received her BA and MA in Sociology from the University of Western Ontario, and her PhD in Human Geography from the University of St Andrews. Her research focuses on the integration and assimilation patterns of Christian Palestinians in Britain, and the transnational experiences of Palestinians in exile. She is currently a Lecturer of Sociology at the University of Derby.

Shail Shankar is an Assistant Professor at the Indian Institute of Technology, Mandi. He received his PhD in psychology from the University of Allahabad. His research work focused on the religious beliefs of the elderly (*Mumukshus*) and how this related to their health status. After completion of his PhD, he joined the *Centre of Behavioural and Cognitive Sciences* (Allahabad) as a post-doctoral research fellow and researched the experiences of pilgrims (*Kalpawasis*) attending the Magh Mela at Allahabad. Currently, he is working on issues of identity amongst Tibetan refugees in India. His research interests include identity and religious beliefs and their linkages to health and well-being.

Clifford Stevenson is a Reader in Psychology at Anglia Ruskin University. His research interests fall in the area of social and political psychology and are concerned with how social identity processes underpin collective expressions of identity, collective resilience and intergroup interactions. Work to date spans the study of collective displays of national identity in crowds and parades, community identity and resilience in disadvantaged urban areas, intergroup contact and cross-community identities in divided societies.

Alan Terry is Senior Lecturer in Human Geography in the University of the West of England. His main research interests are concerned with rural development, participatory development processes and food and water security in southern Africa. He has also worked on community-developed indicator projects in India and South Africa.

Shruti Tewari obtained her PhD in Health Psychology at the University of Allahabad researching achievement and coping with physical disability. She then took up a post-doctoral research fellowship at the *Centre of Behavioural and Cognitive Sciences* (CBCS), Allahabad University. During this time she was responsible for organising and delivering several large-scale surveys of pilgrims (*Kalpwasis*) attending the Prayag Magh Mela. Her interests concern the social psychological determinants of health and well-being. She also has interests in social identity processes and how we process information. Currently, she is working on social identity, attention and coping with stress.

James Thurgill is an Associate Lecturer of University of the Arts, London, where he teaches Media Communications and Visual Culture. James holds a PhD in Cultural Geography from Royal Holloway, University of London, where he completed his doctoral thesis Enchanted Geographies: Experiences of Place in Contemporary British Landscape Mysticism. Prior to this, James taught Media and Cultural Studies at the University of Northampton before obtaining an MA in Cultural Studies from Goldsmiths and later a PGCE from the Institute of Education. James was co-organiser of Uncanny Landscapes (2013), a week-long series of symposiums, workshops, an exhibition and conference that sought to explore the relationships between place, marginality and the strange.

Nina Vollenbröker holds a diploma in architecture (2001) and a PhD in architectural history and theory (2013; funded by the Arts and Humanities Research Council). She is a teaching fellow at the Bartlett School of Architecture, University College London. Nina's research and academic teaching focus on spatialisations of home and belonging in people who do not have a long-term attachment to a single geographical location. She is especially interested in the nineteenth-century trans-Mississippi West and has particular expertise in American travel diaries, photographs and needlework.

Acknowledgements

Our thanks go to Jan Mosedale and Caroline Scarles for encouraging us to put this volume together and to Katy Crossan and Tricia Craggs at Ashgate for steering it through to publication.We would like to thank the contributors for working with us on this project and meeting all of our deadlines, and Tim Edensor for skilfully summarising the points and arguments within the individual chapters in his Afterword.

Special thanks are due to Dr Sarah Evans who has, with grace and good humour, provided efficient and effective technical editorial support in the preparation of this volume. We also wish to express our appreciation to Dr Aga Szewczyk for collating the final manuscript.

Finally, our thanks go to the University of the West of England's Department of Geography and Environmental Management for funding the research for our contribution to this volume and the time required to bring the volume as a whole to publication.

The Editors

Chapter 1

Introducing Sacred Mobilities: Journeys of Belief and Belonging

Tim Gale, Avril Maddrell and Alan Terry

Introduction

This opening chapter addresses conceptual and theoretical issues of relevance to the types of journeys that feature in *Sacred Mobilities*, and provides an overview of subsequent chapters in which the relevant narratives, concepts, theories and debates are explored in more detail. It discusses competing notions of the 'sacred', critical readings of the key concepts of liminality and co-presence, and the turn to mobility in the social sciences and what this, in turn, can tell us about these movements and their meanings.

The volume introduced here is concerned with 'sacred mobilities' in their broadest sense, with 'journeys of belief and belonging' – incorporating a spectrum of those movements and travels prompted by religious, more broadly 'spiritual' and secular-sacred practices and priorities. Drawing on the 'New Mobilities Paradigm' (Sheller and Urry, 2006), it offers useful insight to notions of sacred journeys (Coleman and Eade, 2004), where they intersect with people, objects and other things on the move. In turn, it considers how spiritual meanings and practices, however broadly defined, can shed light on the motivations and experiences ascribed to mobilities (Maddrell, 2011, 2013; Maddrell and della Dora, 2013a, 2013b).

The original call for contributions to *Sacred Mobilities* was constructed with a view to capturing a variety of perspectives on, and examples of, journeying to and with the 'sacred', as understood by participants in those journeys *regardless of religion or their beliefs*. Most of the responses we received focussed on spiritual and secular, not institutionalised religious, journeys, and 'belonging' rather than 'belief', tendencies that are reflected in/across the chapters selected for publication. This shows how notions of the sacred have become stretched in recent years – well beyond institutionalised religion and the study thereof.

Defining the Sacred and the 'Blurring' of Boundaries

In his mid-twentieth century treatise *The Sacred and the Profane,* Mircea Eliade (1957) presented a dichotomisation of the sacred and non-sacred. While the sacred related to those spaces and practices associated with the manifestation of the divine or spiritual world, anything outside of this arena was profane and mundane. Eliade argued that these views were broadly held by adherents of most belief systems, but were not necessarily shared by non-believers. This clear distinction between the sacred and the profane both reflected and reinforced those who believed that theophany (the manifestation of a god or the divine) or hierophany (the manifestation of the sacred) was spatially located in fixed points, such as places of revelation, apparition, healing qualities or miraculous powers. This in turn gave rise to particular geographies of the sacred, as individual bounded spaces and as spatial patterns produced by networks of these spaces. Hence, it is possible to map Aztec temples, Hindu sacred river crossings or *tirtha*, and sites reputed for Marian apparitions. Whilst these sites were often attributed with degrees of permeability between the visible material world and invisible, more-than-representational and other-worldly phenomena (see Maddrell and della Dora, 2013a), they were also attributed with impermeable external boundaries (here is holy, there is not); by extension, in this world view, (full) access to the sacred is typically limited to the initiated believer.

However, this model is challenged by theologies which assert the immanence of the divine (for example the animist belief that spirits are present in all creatures, plants, rocks and so on, or the Protestant Christian belief that God can be accessed at any time and in any place). It is also challenged by theologies and spiritualities, and their related epistemologies and ontologies, which recognise the dynamic and relational qualities of beliefs, practices, experiences, identities and ways of being in the world. Thus, in Western thought and practice in the twenty-first century, the sacred not only includes spaces and practices of formal religion, but also those of alternative and self-spirituality (Heelas and Woodhead, 2005) and everything attributed with the highest value and respect (Milton, 2002). Within studies of pilgrimage, itself an iconic sacred journey, these perspectives are exemplified in two key ways. Firstly, in collections such as Reader and Walter's (1993) *Pilgrimage in Popular Culture*, which includes studies of secular journeys to war graves (Walter, 1993) and football stadia (Davie, 1993). Secondly, in the conceptual shift within pilgrimage studies from a primary focus on particular geographical locations such as shrines (for example Turner and Turner, 1978), to analysis of wider embodied-emotional-spiritual-social-spatial relations. This can be seen in Eade and Sallnow's (1991) *Contesting the Sacred*, with its focus on the interrelation of people-place-texts, Dubisch's (1995) practice-centred study of pilgrimage on Tinos, *In a Different Place*, Frey's (1998) phenomenological account of *Pilgrim Stories. On and Off the Road to Santiago*, and Coleman and Eade's (2004) *Reframing Pilgrimage*, with its particular focus on the agential qualities of movement within pilgrim experience and the processes of sacralisation.

Within the social sciences, pilgrimage is sometimes assumed to be associated only with traditional, god-fearing societies, an 'ancestral behaviour' under threat of redundancy from secularisation in Europe, North America and elsewhere (Knox and Hannam, 2014). However, this overlooks a significant body of evidence to the contrary; for example, an estimated 16.6 million people travel for religious purposes in the United States alone (Saltzman, 2010; cited in Olsen, 2014), and 300 million worldwide (National Geographic, 2011; cited in Notermans, 2012). Looking at pilgrimage specifically, 2.3 million pilgrims attended papal events in 2012 (Catholic News Service, 2013; cited in Olsen, 2014), over one million pilgrims undertake the Hajj to Mecca each year, and cyclical pilgrimages such as the Hindu Kumbh Mela attract 10 million pilgrims, representing 'the largest regular assemblages on earth' (Morinis, 1992: 1).

What then do the pilgrimage practices of these varied faith communities have in common? The potentially multi-stranded or competing characteristics of 'pilgrimage' have been variously identified as: (i) a journey to a sacred place; (ii) life as a pilgrimage towards the afterlife; and (iii) an inner journey of contemplation and reflection; (iv) a religious requirement; (v) an act of penance; (vi) a form of petition, notably for healing; (vii) a means of individual or faith community renewal; (viii) a re-inscription of, or search for, meaning (see Digance, 2006; Dyas, 2004; Eade and Sallnow, 1991; Maddrell and della Dora, 2013b; Winkelman and Dubisch, 2005). For anthropologists Turner and Turner (1978), pilgrimage is essentially the practice of journeying to sacred places. A place is deemed to be sacred by being associated with relics, revelations or miracles, and pilgrims to such places experience aspects of 'liminality' – a familiar trope to which we shall return later. The idea of 'life as pilgrimage' features in several contributions to this volume. Here, the notion of pilgrimage as a lifelong project culminating in one's passing into a realm beyond the phenomenal world transcends individual religions and religion *per se*. Pilgrimage as an 'inner journey' may be facilitated by but does not require physical movement as such, but rather moments of stillness and imaginative travel (see the later discussion of mobility/ies and Maddrell (2011, 2013)). Thus it is important to note that the iconic Western image of a long-distance pilgrim is far from representative of all pilgrimage practices, let alone broader sacred mobilities.

The breadth of practices encompassed by the term 'pilgrimage' can, however, present conceptual challenges. In a response to Knox and Hannam's (2014) 'word games', in which they deliberately conflate the concepts of pilgrim and hedonist, Margry argues for the retention of basic distinctions, and pointedly defines contemporary pilgrimage as:

> ... a journey out of religious or spiritual inspiration, consciously undertaken by individuals or groups, to a place or along a track that is regarded as more sacred or salutary than the environment of everyday life, to seek a transcendental encounter with the sacred for the purpose of acquiring spiritual, emotional or physical healing or benefit. (Margry, 2014)

However, as illustrated by this collection and others, any definition of the sacred and its related spaces and practices, needs to be sufficiently inclusive to accommodate the idea that 'what becomes labelled and adopted as sacred by society does not have to pertain to religion [or spirituality]' (Tresidder, 1999: 141). Hence Morinis' broad definition of pilgrimage as 'a journey undertaken by a person in quest of a place or a state that he or she believes to embody a valued ideal' (Morinis, 1992: 4) is adopted here as a framework for exploring a wide range of sacred mobilities, regardless of whether they are grounded in religion, self-spirituality or secular passions. Digance (2006) has argued that a definition of pilgrimage appropriate to the twenty-first century is needed, and we suggest the notion of 'sacred mobilities' as an umbrella concept which accommodates the wide variety of practices discussed here.

The pilgrim-tourist debate has featured prominently in studies of tourism, religion and spirituality. These issues have been rehearsed at length elsewhere (for example, MacCannell, 1976; Graburn, 1983; Stausberg, 2011) and do not need reproducing in full here, yet merit some brief consideration. There is a noticeable tendency in the literature to emphasise what religious pilgrimage and tourism have in common, and to metaphorically compare the latter to the former. Both require leisure time and financial resources; and are condoned – and in some cases valorised – by the society in which they are practiced (Smith, 1989; cited in Holden, 2005). Also, there are many circumstances in which the interests and practices of 'pilgrim' and 'tourist' overlap; for example, pilgrims acting as tourists en route to a shrine – sightseeing, souvenir-hunting and socialising – and heritage/ cultural tourists, hikers and those in search of the spiritual, rediscovering and appropriating traditional routes such as the Camino to Santiago, or investing a sense of the sacred in more modern and seemingly secular sites like Elvis Presley's home and place of burial at Graceland. This echoes Turner and Turner's (1978: 20) phrase that 'a tourist is half a pilgrim, if a pilgrim is half a tourist'. Extending the comparison further, tourism is oft-described as a 'sacred journey' (Graburn, 1989; Sharpley and Sundaram, 2005), characterised by ritualised practices (such as sightseeing) and sacralised places (such as tourist attractions) (see MacCannell, 1976). Graburn (1983; cited in Sharpley, 2008) goes one step further to argue that they are one and the same: 'if tourism has the quality of a leisure ritual that takes place outside of everyday life and involves travel, is it not identical to pilgrimage?' Indeed, if we accept that the function of holidays (or holy days) in modern society is akin to that of religious festivals in traditional societies, then it is possible to argue that tourism is the modern metamorphosis of pilgrimage, robbed of its symbolic and mystical power by secularisation (Cohen, 1992). Furthermore, it can be argued that pilgrims clearly satisfy the definition of tourists as those who travel from home on a temporary basis for non-work-related purposes. Regardless of whether one subscribes to these views, tourism and pilgrimage share common attributes such as (some) geographical distance from home and freedom, for a time at least, from some everyday responsibilities. Ultimately pilgrims and tourists may have much in common, but their specific motivations and experiences need to be interrogated in order to understand any sacred attributes of that experience. Subsequent references in this chapter to pilgrims and tourists as categories of/ labels attached to people on the move are not *necessarily* mutually exclusive, but serve to advance or frame a particular point or argument that we are making, rather than to assert that they are somehow different and should always be described thus.

Key Concepts: Liminality and Co-presence

Turner and Turner (1978) represented pilgrimage as a rite of passage, involving the spatial (actual) and psychological (perceived) separation from everyday life and the rules and requirements that govern our behaviour, and the crossing of a threshold into 'liminal' time-space (a period of 'neither here nor there', an 'in-between' place). Here, they argued, normal social structures are broken down, and one can encounter the true self in moments of 'flow' and experience a sense of communitas (a shared understanding, or communion) with fellow participants in a heightened state of consciousness, before returning to society transformed by the experience (and often to a higher status). While this idealised model of pilgrimage, and especially the notion of communitas, has been critiqued (for example see Eade, 2000), this process of transformation has been effectively applied to a variety of touristic spaces and settings, notably beaches, hotels, planned events and festivals, even virtual worlds (see Gale, 2009; Getz, 2010; Preston-Whyte, 2004; Pritchard and Morgan, 2006). These may more accurately be described as having 'liminoid' characteristics (after Turner and Turner, 1978) – a self-imposed rite of passage that is not quite liminal or life-changing, but still offers a means of escape from daily norms and routines and the possibility of fulfilment/self-actualisation (if not transcendence). Frequently implicit in liminal (or liminoid) phenomena and processes is a sense of co-presence, of a shared experience that generates a range of emotions, feelings and affects. Hence Urry (2002: 256) argues that 'one should investigate not only physical and immediate presence, but also the socialities involved in occasional co-presence, imagined co-presence and virtual co-presence'. Clearly, religious pilgrimages and other forms of sacred mobilities, including secular-sacred journey-assemblages, represent such time-spaces of co-presence (real, imagined or virtual).

Co-presence also explains another aspect of why people undertake sacred journeys. One of the key distinguishing factors between pilgrimage and tourism is the *motivation* for travel (Vukonic, 1996) although, as noted above, it has to be recognised that pilgrims can 'take time out' to be tourists and that tourists may find themselves unexpectedly experiencing the spiritual (Stausberg, 2011). It may be an obvious point to make, but religious pilgrims travel for reasons of their religion: for spiritual renewal or healing; to seek enlightenment; to enact penance for their sins; or to fulfil a promise they made to themselves or others. This has prompted some commentators to argue that tourists travel because 'tourism is their religion' (Runcie, 1988; cited in Urry, 1990), whereby tourists go in search of novelty and stimulation, or to escape to an environment that permits rest and relaxation (Iso-Ahola, 1982). Regardless of why people embark on sacred journeys (or journeys that for them are 'sacred'), they are in most cases willing participants who commit resources to making this happen (free time, discretionary income). However, there are certain 'obligations' of co-presence which suggest that journeying *per se* is not necessarily an entirely voluntary activity:

- To work, to attend family events, to visit public institutions
- To meet face to face, to develop trust, to note body language
- To spend quality time with family, partners or friends
- To sense a place directly, through embodied experiences
- To experience a particular live sporting, political or cultural event
- To work on objects that have a particular physical location.
  (Adapted from Urry, 2002.)

Some of these obligations relate to (paid) employment, and may be dispensed with here. Others, however, are salient to this discussion. For instance, whatever potential we might ascribe to virtual forms of pilgrimage and tourism as replacements or substitutes for the real thing (see below), most feel sacred journeys require physical proximity to and sensory engagement with sites, routes and objects (to sense and interpret anything that might give meaning to the journey), and also bodily presence(s) (to be seen by others and make a symbolic expression of belief or commitment, and for intense moments of communitas). In this book, we see the compulsion to bathe in a holy river (Chapter 2), to tie ribbons to the branches of an ancient yew tree (Chapter 4), to get as close as possible to the action in a motorcycle road race (Chapter 5), to gaze upon and wonder at crop circles (Chapter 6), to enter a cathedral and celebrate mass at the end of a long journey (Chapter 8), and to visit an erstwhile family home (Chapter 11). These journeys are mostly undertaken in the company of or culminate in meetings with friends and family, colleagues or strangers (fellow worshippers, riders and spectators, artists and performers, tourists, workers, and so on), and generate a range of emotional responses and felt effects such as renewal or loss (sometimes serving to unite people and other times to reflect and reinforce divisions).

The discussion above exemplifies the complexities and apparent contradictions inherent in defining what is sacred and more particularly sacred places and sacred journeys, including those subsumed under the heading of 'pilgrimage'. These include the potential dialectic tensions between outer and inner 'journeys'; collective and individual agendas and experience; a sense of sacred (hierophanic) qualities being found in particular locations and wider beliefs and theologies; and the roles of motion, travel and the extraordinary versus everyday lived beliefs and practices. Thus any individual or collective sacred mobility needs to be seen as an experiential assemblage. Deploying the concept of 'sacred mobilities' in this volume is grounded in recognition of those assemblage qualities and requires an understanding not only of the sacred, but also of mobilities, to which we now turn.

Mobilities and the Mobility 'Turn'

Movement, however we choose to describe it, is full of meaning (Cresswell, 2006), from a walk in the park to a round-the-world trip. Any journey, sacred or otherwise, means something to the participant and will also have significance beyond the level of the individual (even if it may seem trivial). Journeys can be seen to encompass multiple and intersecting mobilities of various individuals (for example couriers, commuters and tourists), and other non-human sentient life forms and non-sentient buildings, objects and machines. Accordingly, and despite appearances, (sacred) sites and routes are not fixed and given but arise from contingent conjunctions of these agents and practices at particular points in space and time; these conjunctions afford the realisation of certain performances that are recognisable as 'commuting', 'pilgrimage', 'tourism' and so on. To use the nomenclature of the social sciences, 'scapes' (complex, but predictable networks) enable 'flows' (unpredictable movements), albeit in a circumstantial rather than deterministic fashion. Crucial to this is how mobility systems distribute people, activities and objects in/through time and space (Mavrič and Urry, 2009). In the Western world, many of these systems originated in the Industrial Age (for example inland waterways, the national postal service, guidebooks, timetables, organised excursions and department stores), whereas others have emerged more recently (motorways and high speed rail networks, jet aircraft, the Internet and so on). However, a mobilities approach goes beyond these systems to consider the experiential and performative dimensions of these movements, their meanings and significance. For example, movement, travel and related practices reflect wider social relations and produce particular rhythms (Edensor, 2010) that shape the place-temporality of landscapes and urban spaces (Wunderlich, 2010), including those shaped by sacred mobilities (Maddrell, 2011, 2013).

Mobility/ies is more than a concept; it is a post-disciplinary paradigm that integrates leisure and tourism, transport and migration, embodied practice and performance, with the potential for new ways of seeing and thinking about these things and the links between them. It is broadly concerned with mobile practices (for example walking, running, driving, cycling), spaces (roads, railways, airports, cities, the internet), and subject positions (tourist, commuter, migrant worker, refugee), and mobility at a variety of scales – from the global to the local (Cresswell and Merriman, 2011; Hannam *et al.*, 2006). Introduced in Sheller and Urry's (2006) ground-breaking paper as the 'New Mobilities Paradigm' (NMP), it can be traced back to earlier work in sociology and geography that problematised societies and regions as the best contexts for research in these disciplines (see Urry, 2000, 2002; Hall, 2005). The NMP calls into question the use of 'sedentarist' theories of people and place to understand a world where everyone and everything is seemingly on the move, travelling further, faster and more frequently than ever before. Not without its discontents (for example, McAllister, 2011), it nevertheless has the potential to transform conventional, discipline-based understandings of sacred journeys in six key ways.

Firstly, there is a move towards 'de-exoticising' travel, so that we are no longer preoccupied primarily with journeys to distant lands. Instead, attention is redirected from reductive descriptions of static places (for example, towns, shrines, places of worship) and predictable people (devout pilgrims, hedonistic tourists, hospitable locals and so on), to representations, practices and performances. Outcomes from this include a recognition of the complex and dynamic positionalities of individuals (for example, tour guide-cum-pilgrim, spiritual sports fan), and an interest in less obvious sacred sites and journeys. By 'obvious', we are not just thinking of, say, Lourdes or the Hajj, but also sites/sights of formal, religious pilgrimage that are not so widely known or celebrated; 'less obvious', therefore, includes spaces of secular pilgrimage and experiences such as dance clubs and raves, which correspond to earlier descriptions of ritualised behaviour and liminal spaces and bodies, with DJs (disc jockeys) taking on the role of 'high priests' on this occasion (Jaimangal-Jones *et al.*, 2010). In this book, David Crouch comes closest to what has been described here with his account of sacred journeys in/of the mind, rather than through space (Chapter 3).

Secondly, processes that have enhanced the personal mobility of some people(s) are understood to heighten, and serve to highlight, the immobility of others (that is, marginalised or subjugated populations or groups within a given population). Opportunities to travel are frequently constrained by inequalities such as poverty, but can also be linked to injustices and broader conflicts in the past and present. This is particularly evident in Suha Shakkour's poignant account of journeys undertaken by exiled Palestinians to (the sites of) former homes now under occupation following the Arab-Israeli wars of 1948 and 1967 (Chapter 11), and also in Katy Beinart's reflections on, and playful re-enactment of, the forced migration of her Jewish great-grandparents in the early 1900s to escape the pogroms of Eastern Europe (Chapter 10).

Thirdly, the NMP accommodates virtual and communicative travel, thanks to various digital, networked devices that can allow virtual participation in real time and negate the need for physical presence. This is demonstrated in studies of virtual pilgrimage (Rosander, 2004) or 'cyberpilgrimage' (Williams, 2013), virtual presence at funeral rites via webcam links (Maddrell, 2012) and virtual worlds as tourist destinations (Gale, 2009). Virtual mobilities are not discussed in detail in this collection, but the role of websites in the rhetorical shaping of sporting events as 'sacred' is discussed by Maddrell *et al.* (Chapter 5). Furthermore, non-corporeal mobility extends to imaginative travel as well; for example, the anticipation and recollection of sacred journeys (to be) undertaken by oneself and others including, in the latter category, people who are no longer alive but whose memory lives on in written records and artefacts. A case in point is Nina Vollenbröker's use of manuscript diaries and patchwork quilts left behind by nineteenth-century settlers in the American West to retrace the journeys they undertook from their original homes in the eastern states (Chapter 9).

Fourthly, there is more (explicit) consideration of the *outcomes* of being on the move and those impacted by them, and of the undesirable and unintended consequences of living mobile lives that are outside our control and threaten that very mobility (for example, dangerous climate change, dwindling oil supplies and terrorist atrocities). The latter represents an important agenda for mobilities studies, but is beyond the scope of this book; however, consideration of outcomes and impacts threads through a number of chapters. For example, many sacred journeys involve an element of risk; these risks can be managed or avoided, but they may also be embraced because they enable participants to apprehend their own mortality and to think about what lies beyond. Our study of the Tourist Trophy (TT) races on the Isle of Man serves to demonstrate this, with respect to those visitors to the Island who attempt to emulate their TT heroes and the speed at which they travel by riding the course on their motorcycles when not in use for races or practice sessions (Chapter 5). In addition to risks associated with mobility, there are mobile risks; these include the risk of death from crushing in mass gatherings and the spread of communicable diseases, as outlined by Nick Hopkins and colleagues in relation to the Kumbh Mela (Chapter 2).

A fifth characteristic, which has emerged more recently in the mobilities literature, is that of 'moorings' (Hannam *et al.*, 2006). This idea recognises that mobilities are not only about movement and the experiences and meanings associated with that movement, but that these in turn relate to, and are dependent upon, particular fixed points, whether transport nodes, destinations, home or 'family'. We can identify traditional notions of sacred places as nodes in pilgrimage and other sacred networks, but as this collection shows (notably in the chapters by Shakkour, Beinart and Vollenbröker) ideas of family, home, community, dwelling, individual and national identity all act as important 'moorings' within sacred mobilities. An individual or community's relation to a particular 'mooring' may be fixed or dynamic and shifting, but understanding moorings is crucial to understanding mobilities, not least sacred mobilities e.g. the importance of shrines or sacred routes.

Finally, the NMP has foregrounded the use of interdisciplinary 'mobile' methodologies and methods in social science research, in order to capture life on the move. These include participant observation, time-space diaries, cartography and geographical information systems (GIS), cyber-research, visual and literary sources, oral histories and so on (a number of which are deployed in this collection). In the context of interdisciplinary pilgrimage studies, this rich methodological toolkit has facilitated nuanced accounts of pilgrim experience. For example, recent detailed studies on working class (Harris, 2013) and migrant women's pilgrimage (Jansen, 2012; Notermans, 2012) supports Clifford's (1997) assertion that pilgrimage is less gender and class-bound, and less Eurocentric, than other forms of travel.

Broadly conceived, this innovative collection of theoretical reviews and empirical case studies goes beyond simplistic analyses of and arguments about pilgrimage and tourism (often represented as the two 'ideal' types of sacred journey), to creatively explore the intersections of travel and both seeking and practising the 'sacred' through sacred mobilities.

Organisation of the Book

Sacred Mobilities is divided into three sections or themes: the Sacred-secular and the Secular-sacred (Section One); Tracing Historical Footprints (Section Two); and Sacred Journey: Home, Family and Nation (Section Three). The narratives, case studies and analyses presented by these individual chapters are subsequently drawn together and discussed in Tim Edensor's Afterword.

The Sacred-secular and the Secular-sacred

Our exploration of the parallels between (and blurring of) the sacred and the secular begins with Chapter 2, by Nick Hopkins, Clifford Stevenson, Shail Shankar, Kavita Pandey, Sammyh Khan and Shruti Tewari. Their study of Hindu pilgrimage draws on social psychological research on crowds and crowd behaviour at more 'mundane' mass events (football matches, national festivals, protest marches and so on), with a view to understanding pilgrims' experience of the Magh Mela at Prayag (Allahabad, India). There are several aspects to this that correspond with the earlier description of the mobilities perspective and what this brings to studies of sacred journeys: it rejects the idea that pilgrimage and tourism are marginal to everyday life and cannot be understood using theories thereof; it draws attention to the many different types of people that come together to perform the Mela and the roles they play (for example, pilgrim, beggar, vendor); it emphasises the acts of movement and bodily performances they engage in (walking, singing and so on); and it highlights the tensions and negative aspects of co-presence, as well as the benefits and transformative properties thereof. It also uses mobile methods (multi-sited ethnography, interviewing on the move).

In contrast, David Crouch invites us to look *beyond* the spaces and rituals of formal religion for the 'sacred' in Chapter 3, arguing that this resides within 'us' regardless of what we believe in (or not). He explores moments and feelings of immanence, of encountering the divine in everyday life and not as a transcendental 'Other' that we feel closer to by undertaking a physical journey to a faraway place or along a route that is deemed worthy of pilgrimage by a powerful religious group. Accordingly, and in keeping with his Quaker beliefs, our lives are to be lived as a personal act of 'pilgrimage' in which we need to be aware of and responsive to the spirituality and wonder that can reside in the most mundane of things, and in day-to-day practices and movements (for example, gardening, walking the dog).

Alternative spiritualities ('alternative' to the world's major religions, that is) also feature in Chapter 4, in which James Thurgill provides an auto-ethnographic account of a neo-Pagan ritual that took place at the Ankerwycke Yew in the Middlesex village of Wraysbury, UK, on the occasion of Halloween 2011. He offers an insight into 'magick' (not to be confused with magic, as a form of entertainment), which has largely escaped scholarly attention to date. The idea that magick is innate in (certain) objects and realised through ritual performance resonates with particular forms of realism (see Gale and Botterill, 2005); indeed, the chapter draws attention to the role of people as active agents in creating and performing ritual (liminal) space, together with the enduring structures and properties of objects that enable and constrain human agency.

Returning to the idea of 'pilgrimage as metaphor', this features in the accounts given by spectators at the annual TT (Tourist Trophy) motorcycle races in the Isle of Man, reported in Chapter 5. Here, we (with our colleague Simon Arlidge) consider how the TT is represented and experienced as a sports pilgrimage, as something that every motorcycle and road racing enthusiast should do 'at least once in a lifetime' (invoking comparisons with Muslim pilgrims journeying to Mecca for the Hajj). This is achieved through a combination of a survey of visitors and local residents, interviews with state and church representatives, and participant observation, further adding to the list of methods used in this volume.

Tracing Historical Footprints

History is foregrounded in the chapters that comprise the second or middle section of the book; for example, in discussing the antecedents to contemporary, post-/secular sacred journeys (for example, legend trips), or the origins and historical development of particular routes or traditions of pilgrimage. In Chapter 6, Rob Irving considers how places are shaped by myths and legends to become 'legend landscapes', a fusion of the psychic and the physical, a place where something extraordinary is said to have occurred. He proposes a typology of visitors to such places of enchantment: religious devotees; New Age mystical tourists; and legend trippers. Marian phenomena or apparitions of the Blessed Virgin Mary are discussed in some detail, and deconstructed, as examples of the melding of religious pilgrimage and legend tripping. However, the emphasis in this chapter is very much on the second of these visitor types, so-called 'New Agers', and their performative activity in and around Avebury, Wiltshire (UK). Designated a World Heritage Site (along with nearby Stonehenge), on the basis of the prehistoric monuments from the Neolithic and Bronze Ages that are to be found there, it is the association of the area with the more recent phenomenon of crop circles that is of interest here. Created by people, or 'croppies', yet deliberately construed as otherworldly events and the impressions they leave behind, crop circles are a demonstration of the multiple mobilities of ideas, behaviours, feelings and experiences that inhabit and give meaning to a legend landscape, according to Irving.

The other two chapters that make up this section of the book focus on the paths along which pilgrims travel or have travelled in the past (in a literal, rather than metaphorical, sense). Pamela Richardson's biography of George Fox and his travels in North West England (Chapter 7) contains some evocative descriptions of the landscapes and people of the region, and brings to life the peripatetic existence of the preacher with its attendant hardships. The chapter moves through a number of different spaces which were central to the development of early Quakerism (the name given to the movement he founded) as a radical alternative religion, from the wild and lonely places through which Fox walked and in which he preached, to the dank dungeons in which many early Quakers were imprisoned, and finally to the meeting houses which have become themselves the objects of modern-day Quaker pilgrimage. The chapter also considers Fox's legacy and the development of Quakerism after Fox, including a number of early Quaker women whose mobility was closely connected to the practice of their faith. The simplicity of Quaker pilgrimage stands in contrast to other, more orthodox sacred journeys (echoing some of the ideas and arguments presented in Chapter 3).

Another such journey is The Way of St James, the restoration of which is described by Belén Castro Fernández in Chapter 8. Drawing on a wealth of Spanish-language literature, the chapter discusses the beginnings of the Way and its rise to medieval glory, and steady decline thereafter until the modern period. It then examines the appropriation of the Way for political purposes, initially under the Franco dictatorship, and how this gradually evolved into heritage/cultural tourism, transcending and perhaps forgetting its religious origins. In its discussion of the contestation of this transition, the chapter provides an opportunity to revisit the pilgrim-tourist debate discussed above, and presents arguments for not conflating these two subject positions, but rather continuing to observe the subtle nuances and distinctions between them.

Sacred Journey to Home, Family and Nation

The sacred journeys reported in the third and final section of the book are inspired by a search for 'home' or some other marker of identity (family, nationality). Whether searching for a new home (Chapter 9) or an old one (Chapter 11), many of the properties of pilgrimage described earlier can be discerned in the physical movements of people and other forms of mobility re-/interpreted in these chapters. These discussions of sacred journeys also share an emphasis on the tangible, material objects associated with home.

In Chapter 9, Nina Vollenbröker advances the idea of 'dwelling' in motion through a study of the Overland Trails of nineteenth-century America. Vollenbröker, as noted above, draws on a wealth of surviving traces of these migrations, including manuscript diaries kept by migrants *en route*, poetry composed in response to the landscapes they traversed, and diverse personal artefacts, including the quilts made by migrant women in spare moments on the road. She purposively counters and critiques established conceptions of rootedness – a desired state in Western cultures – as being contingent upon inhabiting a given location, and argues for the possibility of growing sacred roots in a mobile location. The approach adopted in this chapter could be described as innovative (at least in the context of the prior discussion of mobile methods), by using the creative output of (deceased) others as an expression of life on the move, including not only their written echoes but also, in examining the quilts, their more material traces.

Creativity is also a feature of Chapter 10, in which Katy Beinart narrates a 26 day journey from Antwerp to Cape Town by container ship in 2009. Accompanied by her sister Rebecca, an artist, Beinart was seeking to retrace a journey first undertaken by her Jewish great-grandfather and great-grandmother many years previously, in order to escape persecution back home. Throughout the journey, the sisters sought to create and enact their own rituals of remembrance, baking bread with a grape must culture brought from home, and reconstructing family photographs using the original clothes of their relatives. Beinart draws upon discussions of heritage tourism, pilgrimage, mobilities and performative art practice to develop an account of such embodied play as a space of potential transformation, in which attempts at reconnection with and re-enactment of the past are framed in terms of complexity, (deliberate) confusion, and liminality, rather than being intended as a straightforward and sentimentalised direct recreation. The reader is invited to participate in the ongoing narrative, which may lead to many other interpretations of this particular sacred journey that cannot reasonably be anticipated here.

Chapter 11 also draws on themes of materiality, memory, and self-identity. Suha Shakkour deploys the process of site sacralisation described by MacCannell (1976 [1989]), to discuss how the sites of former homes of Palestinians (and extant buildings) in territory now occupied by Jewish-Israelis are endowed by their former inhabitants with a level of significance similar to that ascribed to sacred places. This process operates on both an individual level – focus on the lost home, its spaces and memories, and on the passed-down relics of now-useless keys and deeds – and also as a broader collective experience and narrative of the *Nakba*. Shakkour draws on a series of interviews conducted with Palestinians living in London. These provide vivid and moving accounts of respondents' experiences of visiting these sites, and of the imaginative travel of those who cannot or will not visit them, which Shakkour places in the wider context of 'diaspora tourism'. This makes for a fitting conclusion to the substantive part of the book, the 'loose ends' from which are dealt with in Tim Edensor's afterword (Chapter 12).

References

Clifford, J., 1997. *Routes: Travel and Translation in the Late Twentieth Century.* Cambridge, MA: Harvard University Press.

Cohen, E., 1992. Pilgrimage and tourism: Convergence and divergence. In: A. Morinis, ed., 1992. *Sacred Journeys: The Anthropology of Pilgrimage.* Westport: Greenwood, 47–61.Coleman, S. and Eade, J. eds, 2004. *Reframing Pilgrimage: Cultures in Motion.* London: Routledge.

Cresswell, T., 2006. *On the Move: Mobility in the Modern Western World.* New York: Routledge.

Cresswell, T. and Merriman P., 2011. Introduction. In: T. Cresswell and P. Merriman, eds, 2011. *Geographies of Mobilities: Practices, Spaces, Subjects .* Farnham: Ashgate, 1–18.

Davie G., 1993. You'll never walk alone. The Anfield pilgrimage, In: Reader, I. and Walter, T., eds, 1993. *Pilgrimage in Popular Culture.* Basingstoke: Macmillan.

Digance, J., 2006. Religious and secular pilgrimage: Journeys redolent with meaning. In: Timothy, D.J. and Olsen, D. H., eds, 2006. *Tourism, Religion and Spirituality*. London: Routledge, 36–48.

Dubisch, J., 1995. *In a Different Place. Pilgrimage, Gender, and Politics at a Greek Island Shrine*. Princeton: Princeton University Press.

Dyas, D., 2004. Medieval patterns of pilgrimage: A mirror for today? In: C. Bartholomew and F. Hughes, eds, 2004. *Explorations in a Christian Theology of Pilgrimage*. Aldershot: Ashgate, 92–109.

Eade, J., 2000. Introduction. In: J. Eade and M. J. Sallnow, eds, 2000. *Contesting the sacred: Anthropology of Christian Pilgrimage*. London: Routledge. 2nd ed., ix–xxi.

Eade, J. and M.J. Sallnow, eds, 1991. *Contesting the sacred: Anthropology of Christian Pilgrimage*. London: Routledge.

Edensor, T., ed., 2010. *Geographies of Rhythm. Nature, Place, Mobilities and Bodies*. Farnham: Ashgate.

Eliade, M., 1957. *The Sacred and the Profane. The Nature of Religion*. Translated by W.R. Trask. New York: Harcourt, Brace and World, Inc.

Frey, N., 1998. *Pilgrim Stories. On and Off the Road to Santiago*. Berkeley: UC Press.

Gale, T., 2009. Urban beaches, virtual worlds and 'the end of tourism'. *Mobilities*, 4, 19–138.

Gale, T. and Botterill, D., 2005. A realist agenda for tourist studies, or why destination areas really rise and fall in popularity. *Tourist Studies*, 5(2), 151–74.

Getz, D., 2012. *Event Studies: Theory, Research and Policy for Planned Events*. Oxford: Butterworth-Heinemann. 2nd ed.

Graburn, N., 1989. Tourism: The sacred journey. In: V. Smith, ed., 1989. *Hosts and Guests: The Anthropology of Tourism*. Philadelphia: University of Pennsylvania Press. 2nd ed., 21–36.

Hall, C.M., 2005. *Tourism: Rethinking the Social Science of Mobility*. Harlow: Pearson.

Hannam, K., Sheller, M. and Urry, J., 2006. Mobilities, immobilities and moorings. *Mobilities*, 1(1), 1–22.

Harris, A., 2013. Lourdes and holistic spirituality: contemporary Catholicism, the therapeutic and religious thermalism. *Culture and Religion*, 14(1), 23–43.

Heelas, P. and Woodhead, L., 2005. *The Spiritual Revolution. Why Religion is Giving Way to Spirituality*. Oxford: Blackwell.

Holden, A., 2005. *Tourism Studies and the Social Sciences*. Abingdon: Routledge.

Iso-Ahola, S.E., 1982. Towards a social psychology of tourism motivation: a rejoinder. *Annals of Tourism Research*, 9(3), 256–61.

Jaimangal-Jones, D., Pritchard, A. and Morgan, N., 2010. Going the distance: locating journey, liminality and rites of passage in dance music experiences. *Leisure Studies*, 29(3), 253–68.

Jansen, W., 2012. Old routes, new journeys: reshaping gender, nation and religion in European pilgrimage. In: Jansen, W. and Notermans, C., eds, 2012. *Gender, Nation and Religion in European Pilgrimage*. Farnham: Ashgate, 1–18.

Knox, D. and Hannam, K., 2014. Is the tourist a secular pilgrim or a hedonist in search of pleasure? *Tourism Recreation Research*, 39(2), 236–42.

MacCannell, D., 1976, [1989]. The Tourist: A New Theory of the Leisure Class, New York: Shocken.

Maddrell, A., 2011. 'Praying the Keeills'. Rhythm, meaning and experience on pilgrimage journeys in the Isle of Man. *Landabrefið*, 25: 15–29.

Maddrell, A., 2012. Online memorials: the virtual as the new vernacular. *Bereavement Car* , 31(2), 46–54.

Maddrell, A., 2013. Moving and being moved. More-than-walking and talking on pilgrimage walks in the Manx landscape. *Journal of Culture and Religion*, 14(1), 63–77.

Maddrell, A. and della Dora, V., 2013a. Crossing Surfaces in search of the Holy: Landscape and Liminality in Contemporary Christian Pilgrimage. *Environment and Planning A*, 45, 1105–26.

Maddrell, A. and della Dora, V., 2013b. Editorial: Spaces of renewal. *Culture and Religion: An Interdisciplinary Journal*, 14, 1–7.

Maddrell, A., della Dora V., Sacfi, A. and Walton, H. (2014) *Christian Pilgrimage, Landscape and Heritage: Journeying to the Sacred.* London: Routledge.

Margry, P. J., 2014. Whiskey and pilgrimage: Clearing up commonalities. *Tourism Recreation Research*, 39(2), 243–7.

Mavrič, M. and Urry, J., 2009. Tourism studies and the new mobilities paradigm (NMP). In: T. Jamal and M. Robinson, eds, 2009. *The SAGE Handbook of Tourism Studies*. London: Sage, 645–57.

McAllister, J., 2011. *Stuck Fast: A Critical Analysis of the 'New Mobilities Paradigm'.* MA. University of Auckland. Available at: https://researchspace. auckland.ac.nz/handle/2292/10039 [Accessed February 4,2014].

Milton, K., 2002. *Loving Nature*. London: Routledge.

Morinis, A., 1992. *Sacred Journeys. The Anthropology of Pilgrimage*. Westport: Greenwood Press.

Notermans, C., 2012. Interconnected and gendered mobilities: African migrants on pilgrimage to our Lady of Lourdes in France. In: Jansen, W. and Notermans, C., eds, 2012. *Gender, Nation and Religion in European Pilgrimage*. Farnham: Ashgate, 19–36.

Olsen, D.H., 2014. Metaphors, typologies, secularization, and pilgrim as hedonist: A response. *Tourism Recreation Research*, 39(2), 248–56.

Preston-Whyte, R., 2004. The beach as a liminal space. In: A.A. Lew, C.M. Hall and A.M. Williams, eds, 2004. *A Companion to Tourism*. Oxford: Blackwell, 349–59.

Pritchard, A. and Morgan, N., 2006. Hotel Babylon? Exploring hotels as liminal sites of transition and transgression. *Tourism Management*, 27, 762–72.

Reader, I. and Walter, T., eds, 1993. *Pilgrimage in Popular Culture*. Basingstoke: Macmillan.

Rosander, E.E., 2004. Going and not going to Porokhane. Mourid women and pilgrimage in Senegal and Spain. In: S. Coleman and J. Eade, eds, 2004. *Reframing Pilgrimage: Cultures in Motion*. London: Routledge. 69–90.

Sharpley, R., 2008. *Tourism, Tourists and Society*. Huntingdon: ELM Publications. 4th ed.

Sharpley, R. and Sundaram, P., 2005. Tourism: A sacred journey? The case of Ashram Tourism, India. *International Journal of Tourism Research*, 7, 161–71.

Sheller, M. and Urry, J., 2006. The new mobilities paradigm. *Environment and Planning A*, 38(2), 207–26.

Stausberg, M., 2011. *Religion and Tourism: Crossroads, Destinations and Encounters*. London: Routledge.

Tresidder, R., 1999. Tourism and sacred landscapes. In: D. Crouch, ed., 1999. *Leisure/Tourism Geographies: Practices and Geographical Knowledge*. London: Routledge.

Turner, V.W. and Turner, E., 1978. *Image and Pilgrimage in Christian Culture: Anthropological Perspectives*. New York: Columbia University Press.

Urry, J., 1990. *The Tourist Gaze*. London: Sage.

Urry, J., 2000. *Sociology Beyond Societies: Mobilities for the Twenty-First Century*. London: Routledge.

Urry, J., 2002. Mobility and proximity, *Sociology*, 36(2), 255–74.

Vukonic, B., 1996. *Tourism and Religion*. Oxford: Elsevier.

Walter, T., 1993. War grave pilgrimage. In: I. Reader and T. Walter, eds, 1993. *Pilgrimage in Popular Culture*. Basingstoke: Macmillan, 63–91.

Williams, A.M., 2013. Surfing therapeutic landscapes: Exploring cyberpilgrimage. *Culture and Religion*, 14(1), 78–93.

Winkelman, M., and Dubisch, J., 2005. 'Introduction: The anthropology of pilgrimage'. In: *Pilgrimage and Healing*, J. Dubisch and M.Winkelman, eds, Tucson, AZ: University of Arizona Press, ix–xxxvi.

Wunderlich, F.M., 2010. The aesthetics of place-temporality in everyday urban space: the case of Fitzroy Square. In. T. Edensor, ed., 2010. *Geographies of Rhythm. Nature, Place, Mobilities and Bodies*. Farnham: Ashgate, 45–56.

SECTION I
The Sacred-secular
and the Secular-sacred

Chapter 2

Being Together at the Magh Mela: The Social Psychology of Crowds and Collectivity

Nick Hopkins, Clifford Stevenson, Shail Shankar, Kavita Pandey, Sammyh Khan and Shruti Tewari

Introduction

In 2013, readers of the *Financial Times Weekend Magazine* would have come across an article describing the Kumbh Mela at *Prayag*, Allahabad. The piece conveyed the scale of this north Indian Hindu pilgrimage event and its spiritual significance. It also conveyed something of the psychological experience of participation: The author – a western expert on yoga – described how, surrounded by the flow of holy men sweeping down to the Ganges, he had 'never felt so much at one with a crowd. With everyone stripped down to their loincloths, a sense of individual identity fell away and we surrendered ourselves to the common purpose' (Mallinson, 2013: 25).

Accompanied by photographs of bearded sadhus and descriptions of naked ash-smeared holy men, the author's account of his experience of being subsumed into the mass, is truly fascinating and immediately evokes Turner's concept of *communitas* (Turner, 1973). Indeed, it is hard to imagine how readers of such a piece could fail to be struck by the contrast between the intense emotional experience of this colourful collective event and the altogether greyer reality of a mundane British weekend. If ever an event was exotic, mystical, and removed from everyday life, surely this – the Kumbh Mela – is it.

However, such contrasts can be misleading, implying that our theories of everyday life have no applicability to such events. Furthermore, they can encourage the belief that whatever happens in such mysterious and mystical events is marginal to everyday life. Yet, we will argue otherwise: In what follows we consider how analyses of everyday social psychological processes can shed light on the dynamics structuring participants' experiences in the Mela, and how such experiences may have longer-term consequences.

Hindu Pilgrimage in India

Pilgrimage is important in Hinduism. The *Mahabharata* and the *Puranas* refer to *tirthas* (literally, fords or crossing points in water) as holy places where one can catch sight of the deities. They also talk of '*tirthas* of the heart' 'implying that the pilgrim should not only bathe in the waters of earthly tirthas, but also in the inner virtues of truth, charity, patience and self-control' (Coleman and Elsner, 1995: 149). For some this state is achieved by adopting the ascetic existence of the renunciant. Yet for most, the renunciation of everyday comfort is more temporarily achieved in and through pilgrimage.

There are many holy sites in India (Bhardwaj, 1973), and Singh (2011) cites evidence that of all the domestic travel in India, over a third is related to pilgrimage. The significance of such flows is hard to overestimate. They have important economic and environmental dimensions (Bleie, 2003), and in the recent past have been given a distinctive political dimension through being used to mark out 'Hindu' territory and advance an ethnic definition of the Indian nation (Jaffrelot, 2009; Singh, 2011). They also have important social psychological dimensions, and across disciplines there is increasing interest in the 'inner experience' of pilgrimage (Collins-Kreiner, 2010).

The Prayag Melas

The Kumbh Mela at Allahabad is the most famous Hindu pilgrimage (for its history, see Maclean, 2008; for a contemporary description, see Tully, Lannoy and Mahendra, 2002). Occurring every twelve years, it is attended by millions from across north India. Yet, although it attracts particular attention, this event is but a special incarnation of an equally remarkable annual gathering that takes place in the month of *Magh* (January/February). Each year, for one lunar cycle, a vast tented city grows up on the sandy floodplain at the confluence of the Yamuna and Ganges rivers (*Prayag*, Allahabad). From nothing, roads appear, pontoon bridges are secured over the Ganges, residential camps are marked out, the more celebrated religious organisations pitch huge marquees, and pilgrims arrive (some by bus and train, some by tractor-towed trailers) from across north India (especially Uttar Pradesh, Madhya Pradesh and Bihar). Some pilgrimage sites are remote and the journeys they require take one into sites of natural beauty. Yet, although the Prayag Melas take place just outside the city of Allahabad, many of the distinctive features of pilgrimage are to be found: travel, hardship and people in abundance.

All sorts of people can be found at the Mela. There are the 'day-bathers' who do not stay in the Mela but who visit, especially on the more auspicious days of the lunar calendar, to bathe in the Ganges. There are beggars and those who service the Mela, sweeping the roads, removing rubbish, cleaning latrines, and undertaking a myriad of other tasks. There are barbers, shopkeepers, police officers, and NGOs promoting their messages. Many of those who visit for just a few days will combine their bathing in the Ganges with a visit to a local temple and then proceed to the markets and funfair that link the site to the city.

However, at the event's heart are those who commit to live at the site for the entire month. For these, the markets and funfair are to-be-avoided distractions, and their daily routine involves movement between the camps in which they live, the tents associated with their gurus and sadhus, and their bathing ghats. These pilgrims are mostly rural and elderly – predominantly in their 60s and 70s (Tewari *et al.,* 2012) – and they seek religious merit by renouncing all worldly ways and comforts in order to live a spiritual existence. Moreover, they do not only commit to live a simple life for the month – they commit to attending for 12 consecutive years. Such pilgrims are known as 'Kalpwasis' and according to some accounts this name gives insight into their purpose – *kalpa* denotes the transformation of the self through inner resolve, and *vas* denotes the living out of this resolve.

Figure 2.1   Holy men in a procession inside the Mela site
Source: Authors' own photograph

It is understandable that journalists should be drawn to the more spectacular displays of faith and spirituality apparent at the Prayag Melas: the sights of sadhus in colourful robes or of naked and ash-smeared renunciants are indeed dramatic. However, an exclusive focus on such figures obscures other equally remarkable characters: the ordinary men and women who undertake Kalpwas. A Kalpwasi's life is hardly spectacular. They do not wear the bright robes of the sadhus nor the ash of the renunciants. Rather, they adhere to a simple routine. They live in canvas tents, sleep on straw, eat one meal a day which is made up of bland foods, eschewing anything spicy which might excite the body to the detriment of the spirit or dull the mind (so-called '*tamasic*' foods). Their days start early – well before dawn they rise to perfom their *puja* before walking to the Ganges for the most important part of their rituals – bathing in the sacred river. On returning to their camps, they devote themselves to the mundane tasks of camp life: fetching water from the standpipes, preparing the midday meal, buying wood or dung for the stoves on which the meal will be cooked, talking with their neighbours. Much of their remaining time is spent on reading holy texts or attending afternoon meetings held in tented venues across the site. Many venues are small – with room for just a few dozen to sit on the floor – but others accommodate several hundred. Some venues host performances of scenes from the holy scriptures and the singing of religious songs. Others host sadhus (typically men) delivering injunctions on religious duties and obligations. Still others host meetings on social issues (for example, the pollution of India's holy rivers).

Camping for a month is hard enough. However, doing so in winter is particularly tough – especially as Kalpwasis must bathe before dawn when the January temperatures before sunrise are not much above freezing. Moreover, the environment is anything but tranquil. The site is peppered with loudspeakers. Some belong to the Mela authorities, others to religious organisations, and such is the multiplicity of competing broadcasts that one is often surrounded by an undecipherable cacophony. Moreover, it is crowded, with limited opportunities for quiet solitude. Indeed, in his account of the Kumbh, Mark Tully draws something of a contrast between the still of the natural world and the busy-ness of the social world of the Mela. Referring to a scene at the water's edge he describes how 'night herons stood silent and still, oblivious of the cacophony and the crowds, their heads jutting forward, waiting to pounce on their prey' (Tully *et al.*, 2002: 58).

Yet, despite all this – the basic conditions, the cold, the noise, the crowding – the Kalpwasis we interviewed described their experiences in overwhelmingly positive terms and as involving a sense of serene bliss (Cassidy *et al.*, 2007; Prayag Magh Mela Research Group (PMMRG, 2007; Shankar *et al.*, 2013). It might be thought that this says more about life in India in general than about life in the Mela itself. Perhaps the harshness of the Mela is not so different from routine village life? Perhaps the noise and the cold are unnoticed? Perhaps people in more 'collectivistic' cultures (such as India) experience others' physical proximity differently from those in more individualistic societies? Yet our ethnographic observation of Kalpwasis in their villages and in the Mela shows that the two environments do indeed differ, and that life at the Mela is difficult. Moreover, we know that the cold and the noise are prominent features of their experience (Shankar *et al.*, 2013; Pandey *et al.*, in press). We also know that people in India can experience crowding as aversive (Nagar and Pandey, 1987).

Yet, still the experience is described in extremely positive terms, and as we will see, this positivity extends to a sense of well-being that lingers even after their return home. What then can social psychology say about this experience and what can it say about the processes involved?

The Experience of Pilgrimage

Pilgrimage sites can be experienced as 'spaces of renewal' (Maddrell and della Dora, 2013a). For example, the spiritual experience of visiting Ireland's holy wells is shaped by the calm and stillness of these sites (Foley, 2013), the rhythm of movement, and the fact that such embodied practices are performed alongside others (Maddrell and della Dora, 2013b). Yet, whilst the quiet and remoteness of a place may invite spiritual calm, at other sites the scale and clamour of the crowds may be particularly important, and as social psychologists we are particularly interested in the collective dimension to the experience of pilgrimage.

The pleasures of participating in collective events have long puzzled observers (Ehrenreich, 2007). Indeed, such is this puzzlement that many analysts have characterised the passion of the collective as indicating an aberrant departure from everyday 'normality'. Certainly, psychological theories have tended to couple emotion and crowd psychology together, and to set them apart from reason and individual psychology. This is clearly apparent in the work of the founding figure of crowd psychology – Gustave Le Bon (1841–1931) – who argued that crowds were characterised by a heightened emotionality and irrationality that could be contrasted from the reason and self-control people exhibit when alone. Indeed, for Le Bon, people in crowds have lost their identity and their capacity for reason such that all that remained were the passions (Le Bon, 1895). Following Le Bon's lead much social psychological theory assumed that participation in mass events entailed the distortion of normal functioning with the consequence that for years, crowds were, as Reicher (2011) puts it, 'the elephant man of the social sciences' (Reicher, 2011: 434).

Outside of psychology, Le Bon's contemporary, Durkheim (1858–1917) offered an analysis in which 'effervescent' assembly was understood in more positive terms – as revivifying social bonds and recharging the collective representations on which everyday social life is based. Durkheim also believed that emotion could overwhelm crowd members, but rejected the idea that people lost their identities in the mass or that the emotions expressed in the mass reflected such losses. As Olaveson (2001) explains, Durkheim's effervescence is 'characterised by intimacy, intensity, and immediacy, yet it also involves will and intention, and symbolic focus', and should not be seen as 'simply mob psychology or camaraderie' (Olaveson, 2001: 101). Moreover, Durkheim argued that the pleasure of the mass arose in part at least from the feeling that people were able to transcend everyday concerns and to enact their shared ideas and values. That is, although in everyday life 'the mind is chiefly preoccupied with utilitarian and individualistic affairs'

(Durkheim, 1912 [1995]: 352; cited in Olaveson, 2001: 110), people in the crowd could rise above such concerns. Durkheim also underlined the importance of the sense of collectivity experienced at the event: he argued that for excitement to occur, a gathering 'must possess a degree of unity, of intimacy and the forces which it releases must be sufficiently intense to take the individual outside himself and to raise him to a superior life' (1913, cited in Olaveson, 2001: 101).

Undoubtedly, Durkheim's approach helps re-habilitate collective events as topics of social enquiry and relevant for everyday life. However, the social scientific theorisation of the psychological processes underlying people's experience of collective participation has been hesitant. For example, Victor Turner's analysis of pilgrimage, although sharing much in common with Durkheim's approach (Olaveson, 2001), presumes that pilgrims' sense of collectivity and heightened emotion arise because participants are freed from everyday structural constraints, and he introduced the term 'communitas' to describe a 'spontaneously generated relationship between levelled and equal total and individuated human beings, stripped of structural attributes' (Turner, 1973: 216). As Ehrenreich (2007) explains, this helped give pilgrimage and the analysis of group behaviour a legitimate place in anthropology. However, as she continiues, there is a sense in which 'it was a marginal and second-rate place' (Ehrenreich, 2007: 11); such collective events were characterised as offering a form of occasional relief from more structured everyday reality.

Moreover, this characterisation of collective events as marginal to the bread and butter issues of social science meant that their analysis was rarely informed by theories of everyday social and psychological processes and slipped instead into a rather mystical celebration of 'anti-structure'. Nor does this emphasis on the liminality of pilgrimage encourage consideration of how the analysis of such events could feed back into helping us understand the nature of society and everyday social processes. However, recent empirical research (typically of crowds in conflict) shows the linkages between crowd events and wider social processes (Reicher, 2011). So too, recent studies of pilgrimage have been increasingly sensitive to how what happens in pilgrimage is bound up with (and illuminates) wider social processes beyond the event itself. Most obviously, recent studies show that a sense of collectivity and joy is not always achieved and that one often finds factionalism and conflict. For example, Messerschmidt and Sharma (1981) identified numerous social divisions amongst Hindu pilgrims attending an event in Nepal, and concluded that if anything 'this pilgrimage is fundamentally and unequivocally a structure-*affirming* occasion' (Messerschmidt and Sharma, 1981: 572, emphasis original). Moreover, and again questioning the characterisation of pilgrimage as liminal to everyday life, researchers have highlighted the degree to which the meanings of a religious site may be contested (Sallnow, 1981: Eade and Sallnow, 1991). Nor is such factionalism and conflict simply the perversion of a fraternalistic ideal. By their nature, religious sites are vessels into which competing meanings may be poured, and wider, everyday collective identities contested

(Coleman, 2002). This is one reason why the Prayag Melas attract politicians and activists from across the political spectrum.

Given the problems associated with conceptualising crowd events in general (and pilgrimage in particular) as beyond the realm of everyday social processes, we believe much may be gained through founding our analysis on social psychological theory developed to analyse everyday group behaviour.

Social Identities

Over the last 40 years social psychological analyses of group behaviour have changed dramatically. The concept of *social identity* developed as an alternative to theories which sought to explain group phenomena in terms of personal and interpersonal processes. Henri Tajfel (1978) drew a distinction between individual-level and group-level processes which was developed by John Turner (1982) and elaborated in what became known as *self-categorisation theory* (SCT) (Turner *et al.*, 1987). This approach to collective behaviour emphasises the role of social categories in self and other construal. It argues that the self and others may be defined at different levels of abstraction. Sometimes this may be in terms of individual uniqueness. At other times it may be in terms of specific social categories that give rise to particular social identities. This shift in the salience of individual-level *personal* and group-based *social* identities is psychologically important and makes group behaviour possible. Specifically, the basis for group formation lies in the cognitive act of self-categorisation: it is when people define themselves in terms of membership of the same social category, rather than their unique individuality, that they begin to act as group members. First, people's behaviour is in accordance with their understanding of their collective identity. Second, people's social relations with one another are changed such that to the degree that they view themselves as sharing a common identity, their interactions are more trustful (Tyler and Blader, 2001), respectful (Renger and Simon, 2011) and helpful (Levine *et al.*, 2005; Wakefield *et al.*, 2011).

In recent years this approach to groups has been applied to crowds. Research shows that crowd behaviour is neither random nor uncontrolled but is in accordance with crowd members' understanding of their common social identity (Reicher, 1984). Moreover, unlike traditional social psychological models, which explain group processes as arising out of inter-personal bonds between individuals, the social identity approach is capable of explaining how erstwhile strangers can come together and act together in large-scale collective action. This explains how people – even if they have never met before – can cohere as fellow Catholics in St Peter's Square, as fellow protestors in Tahir Square, or as fellow Manchester United fans at Old Trafford. Moreover, it is when people self-define in terms of a common category that they feel more comfortable in close proximity with fellow group members (Novelli *et al.*, 2010).

Yet, if a common social identity makes group behaviour possible, a common identification cannot be assumed. How an individual conceptualises themselves is variable and the emergence of a shared social identification, a social accomplishment. The boundaries of belonging (who is/is not included), and what it means to identify as a group member can be contested in argument (Hopkins and Kahani-Hopkins, 2004; 2009) and such arguments are sustained and re-shaped by all manner of social practices. For example, analyses of intergroup conflict at demonstrations show how the indiscriminate policing of an otherwise heterogenous collection of individuals and groups can bring about the emergence of a unified and oppositional crowd identity (Reicher, 1996; Drury and Reicher, 2000). Research also shows that although self-definition is important, one's identity is also bound up with the responses of others: just because one identifies with others does not mean such others will recognise and reciprocate that shared identification (Hopkins, 2011; Joyce *et al.*, 2012), and without such recognition of one's social identification, the experience of that identification (no matter how dearly it is held) may be compromised (Hopkins and Greenwood, 2013).

In sum, this social identity perspective on group processes provides a coherent framework with which to approach all manner of routine group processes (conformity, trust, leadership, the provision of mutual support and so on), and has been shown to be of value in analysing conflictual crowd events. We also believe it can help illuminate aspects of a pilgrim's life, and in what follows we reflect on themes arising from our interviews with 37 Kalpwasis (24 men and 13 women, average age 64) conducted in 2010 and from ethnographic observations taken shortly before, during, and shortly after the 2011 Mela. For details on the interview schedule, and the ethnography (which focused on two different Kalpwasis, resulting in a combined total of 115 days of field notes) see Shankar *et al.* (2013) and Pandey *et al.* (2014). These data were gathered by the third and fourth authors (both Indian Hindus) and analysed by a team led by the first and second authors (white English and Irish respectively).

Shared Identity in the Mela

The Scope of Shared Identification

The crowds at the Mela may appear an undifferentiated mass. However, the self-definitions of those attending the event are complex and the Kalpwasis routinely differentiate themselves from non-Kalpwasis visiting the event for a day or two and who combine their bathing in the Ganges with a visit to the funfair or the markets. For example, Kalpwasis often complained that the latter's behaviour was inappropriate because they did not 'know' how to perform rituals properly or because they had a more touristic ethos and exhibited less considerate behaviour. Thus, one interviewee (P15: male, 73 years) complained:

... guys of a young age come. They run and move, some boatman are also like this, who run and move around. The youngsters do this – they do not realize that some old man might fall from their [carelessness]. But the Kalpwasis who are there, they do not do this [Interviewer: They do not?] They take care of this thing.

Another (P6: male, 53 years), argued that different sorts of people attended the Mela and that 'some come for pleasure'. Using a metaphor of a hand, he continued: 'all the fingers are not alike, some people are here for fun', and continued 'but there are no Kalpwasis, no Kalpwasis like that ... [Kalpwasis] all help each other, nothing like that! The people who come, come for bathing. Outside people come, so sometimes, some people come for fun, so sometimes they push and like that'.

Kalpwasis also observed that the presence of these others could limit their own mobilities and their freedom to bathe as they wished. So, when asked if there were differences between Kalpwasis and others 'who come here only for bathing', one interviewee (P21: female, 60 years) replied 'There is quite a lot of difference. Actually it creates trouble. Like when there is more crowd, the police will put barrier[s] on the way while coming and going'. This interrupted the Kalpwasis' routines: 'The police will stop. They don't understand'. Worse, some complained that amongst the non-Kalpwasis there were thieves or young men whose only purpose in bathing was to sneak a look at female bathers. In turn, several reported avoiding areas of the Mela site that were frequented by non-Kalpwasis. Thus, even the Mela site was socially differentiated with consequences for the Kalpwasis' mobility and their spiritual experience of place.

The Psychological Significance of Shared Identity

If Kalpwasis often reported a psychological distance from non-Kalpwasis at the Mela, they also reported a sense of commonality with other Kalpwasis. This could be expressed in various ways. It often entailed references to family-based metaphors (see too PMMRG, 2007). For example, one (P 34: male, 55 years) explained that 'each Kalpwasi, for this one month, feels that we are in our home with our family'. Another (P 11: male, 72 years), explained that when Kalpwasi met each other they would 'do *pranam*' (that is a gesture of respect) and that 'here, the Kalpwasi family is one!' Moreover, they continued: 'Kalpwasis are a family, all with each other! After getting out from here, there, somewhere in the market, people who have come will meet, will do *pranam* and would talk, how was the time spent ... Here the Kalpwasis are family! Everybody is a saint'.

Here there is a clear sense of a shared social identification as Kalpwasis. This does not simply mean that one identifies with a given social category. Nor does it mean that oneself and others identify with the same social category. Rather, there is a sense that crowd members view themselves and each-other in terms of this common social category (Neville and Reicher, 2011), and in this extract this is signaled by the mutual giving and receiving of a greeting (*pranam*) which conveys heightened respect (for a discussion of the role of normatively prescribed greetings in communicating the mutual recognition of identities, see Hopkins and Greenwood, 2013).

A shared identity also entails a sense of common purpose. This was often most conveyed in Kalpwasis' reflections on how the crowds at the Mela compared with those elsewhere, for example at Indian railway stations. One (P10: male 55 years) explained that the Kalpwasi crowds constituted 'some sort of cultural crowd, you can call it a cultural crowd. You can't sleep all night, loudspeakers are blaring, there is noise. People are singing Lord Ram's name. The reciting of the Ramayana is going on. There are religious songs going on. All these things'. He then turned to the crowds at the railway station and observed:

> ... that crowd is not social, it is personal. This [the Kalpwasis crowd] is a social crowd. You can call it a social and cultural crowd. This is the difference. That [the railway crowd] is the crowd of necessity. Since people are passing through, there is a crowd. Here, people stay put, we have to stay for one month, sing religious songs, *Sita-Rama, Sita-Rama, Om Namah Shivay*, this is what we have to do.

Here there is a clear contrast between what is essentially an aggregate of individuals each with their different interests and goals (the railway station) and a collective entity with a common identity and purpose (Kalpwasis as the Mela).

Furthermore, in contexts where others are construed as fellows (such that there is a psychological connection or communion), their presence adds to one's own experience of the event. Sometimes others' presence was important because it validated beliefs associated with their shared identity as Kalpwasis. Thus one (P30 male, 40 years), using the unit of a *lakh* (100,000) and referring to *darshan* (glimpsing the divine), explained that the presence of the crowd showed that:

> I, alone, am not the only one associated with this *dharma*, *lakhs* of people have faith, they are seekers of *darshan*! [Interviewer: OK] This tells us that why the crowd comes here, I alone do not come here because the purpose for which I have come ... a crowd of *lakhs* of people have come and their purpose is the same.

As a corollary, a lack of a crowd may be greeted with disappointment (and not relief), and this respondent explained 'if there was no crowd, so what will someone do? ... The meaning of Mela is the crowd. If there is no crowd, the Mela would feel bad [Interviewer: OK] the Mela won't feel right'.

Shared Identity and the Co-ordination of Behaviour

The wider significance of a shared identity amongst Kalpwasis is that it facilitates individuals' attempts to realise the behaviours associated with their Kalpwasi identity. Most obviously, a shared identity aligns individuals' intentions and

Figure 2.2   Bathers on the river bank
Source: Authors' own photograph

behaviours and such an alignment is psychologically and socially important. For example, the Kalpwasi social identity entails seeking to relinquish worldly concerns so that one can devote oneself to spiritual matters (especially bathing and prayer rituals). This requires considerable individual effort. However, it is not simply an individual achievement. As Kalpwasis live cheek by jowl alongside each other the behaviour of one's fellows is not incidental to one's ability to realize and enact the values and ideals of the Kalpwasi identity. Thus, if one is to avoid the distractions of music or argument, it is important that music and angry voices do not intrude into one's environment. Individual Kalpwasis can decide whether to visit the more commercial or touristy areas of the Mela site (and typically avoided doing so, in part because the non-religious noise associated with such areas is judged as intrusive and as incongruent with their Kalpwasi identity). However, an individual Kalpwasi cannot avoid their fellow Kalpwasis, and their shared commitment to identity-related injunctions is crucial to the ability of any one individual to realise those injunctions.

Similar issues arise in relation to the injunction that one should not gossip. If this goal is to be realised it is important that others share one's identity-related commitments, for without this, one's own ability to live up to this identity-related ideal could easily be compromised. Again, the point is that a shared identity is important because it facilitates one's own ability to enact one's Kalpwasi identity. This is well-illustrated in the following exchange in which the interviewee refers to the concept of *satsang* which translates roughly as an assembly of persons who listen to and speak spiritual truths. First, the interviewee (P37: female, 62 years) stressed the importance of experiencing this *satsang*. Second, she then continued to explain that a key element in achieving this was to avoid mundane gossiping about others' failings. Thus, she explained that:

> ... the most important thing here is the *satsang*. No one gossips about others. No one wants or looks for failings in others ... All become like one family. This is what is called *satsang*. Kalpwas means this only – that you do not criticize or gossip about each other. Each one follows the rules.

Here there is a clear sense in which the commonality of purpose associated with the sharing of a collective identity facilitates an individual's ability to enact group-related ideals and injunctions. In turn, it becomes clear just why the experience of a shared identity can be so rewarding: others' behaviour helps build or scaffold the social context in which one can more easily express one's Kalpwasi identity. Indeed, this same interviewee continued to explain that it was through being able to 'follow the rules' (for example, avoiding gossip and other worldly activities) that they could achieve their spiritual goals such that 'for one month we see heaven, after that daughter, [we are] again in to that same *mayajaal*' (a term referring to the trap of worldly affairs).

This social scaffolding can take diverse forms. A shared identity encourages people to act in terms of a common understanding of identity-appropriate behaviour, and some of these forms of behaviour may add to one's spiritual experience. Some activities (for example, walking, singing, chanting) can result in an 'embodied-sensory-affective-spiritual experience' of the environment (Maddrell and della Dora, 2013b). Moreover, the fact that one participates with others in such performances contributes to this experience. Thus, when asked about her experience of the crowd, one (P25: female, 60 years) explained:

> We feel good. All move singing and playing instruments, the ones who know the song, join them in singing [Interviewer: Yes] we chant, we recite, whatever is on the sound system we join in [Interviewer: Yes] it transforms the mind. We don't even remember our home, where we are. There is so much of noise (laughs).

To sing or chant on one's own is one thing, but to do it collectively is another, and is experienced positively because it helps immerse one in a world where everyday concerns are but distant memories. It should also be noted that such data illustrate the wider point that any assumption that a noisy environment is psychologically draining must be treated with caution: if the noise is construed as identity-affirming it can be experienced positively (Shankar *et al.*, in press; see too Srinivasan *et al.*, 2013).

Shared Identity and Social Support

In addition to facilitating the co-ordination of identity-related behaviour, a shared identity may also be experienced positively because of the belief that if one needs help, support from one's fellows will be available. Certainly, our interviewees reported expecting to receive help should they need it. For example, one (P6: male, 53 years) explained that 'we like the crowd ... there is no question of trouble ... see I feel that the purpose for which I am here, they are also here – everyone is taking dip together'. When asked if these others ever disturbed their rituals, his friend interjected 'no' and explained that there was mutual support. As she put it, 'if we don't have match-stick, we will ask for it from someone and they will give, if they will ask for something from us, we will give', at which point the original interviewee continued 'so everyone helps ... they pay attention on each other, I don't think outside anyone will do this much!'

Of course, it might be that the interviewees are mistaken in their evaluation of the differential likelihood of receiving support from their fellow Kalpwasis and those outside the Mela. However, that is not necessarily the issue. Rather, the point is that these interviewees *expected* support from their fellow Kalpwasis, and the expectation of support can, in and of itself, be important in creating a sense of confidence in going out into and enjoying the world. This confidence is particularly important in an environment such as the Mela. As we have hinted, pursuing everyday routines – particularly when it is so cold – poses considerable difficulties and a shared identity with others can help one manage them such that one can accomplish one's obligations as a Kalpwasi (Pandey *et al.*, 2014). More generally, the supportive relations (even with those whom one does not know) associated with a shared identity facilitates a sense of trusting others. Amongst our interviewees, this was manifested in the belief that others would not harm them (for example, through stealing their property) and several recounted stories in which fellow Kalpwasis ensured that lost items were returned to them. It was also manifested in the belief that those with whom one shared an identity as Kalpwasis would proffer help whenever it was needed, drawing contrasts between what could be expected in the Mela and back home. For example, one (P16: female, 57) explained 'in the home, or in villages, people see each other and feel tensions [Interviewer: OK] Yes. Sometimes because of their earning, sometimes because of looks, sometimes because of fighting. In village it's like this, But here, so much of goodness comes in people, even more than at home'. She continued 'someone will say "come sister sit, warm up your hands and feet" ... It feels good. And there [back home], if you interact more, even in a family, the close relatives cannot stay long with each other!'

The Longer-term Impact of Participation

Thus far we have considered how a shared identity with other Kalpwasis shaped our interviewees' experiences in the Mela. In doing so we employed a theory shown to have value in explaining everyday social and group processes. Applying this to the Mela allows an appreciation of some of the otherwise seemingly inexplicable pleasures of participation. However, the question remains, to what extent does what goes on inside the Mela impact upon life outside the Mela, and more generally, what can we learn about everyday life through studying the Mela in this fashion? Our answer to both these questions involves reference to the issue of shared identity and well-being.

As already hinted, pilgrimage sites can be seen as constituting 'therapeutic landscapes' (Maddrell and della Dora, 2013a) with the spiritual meaning of place contributing to well-being (Kamitsis and Francis, 2013). However, the positive outcomes of pilgrimage cannot be assumed. Mass gatherings bring health risks (Memish *et al.*, 2012). The most obvious concern is crushing (and sadly 36 pilgrims attending the 2013 Kumbh Mela died in one such incident at Allahabad's main railway station) or the potential for spreading communicable disease (such as influenza) (Abubakar *et al.*, 2012). Moreover, the environmental conditions at such events (including crowding and noise) can give rise to health concerns (Steffen *et al.*, 2012), and when events last as long as the Prayag Melas such risks can only be compounded.

Set against this, we also know that feeling socially connected with others can also contribute to well-being. The more that people are involved in social networks – that is, the more social capital they have – the better they fare (Kawachi *et al.*, 2008; Smith and Christakis, 2008), and interventions that increase one's sense of connectedness with others can improve well-being (Walton and Cohen,

2011). Recently, social psychological research has explored just how and why being a member of a group may improve well-being. Research in institutionalised settings (Haslam *et al.*, 2005; Haslam and Reicher, 2006) and in more informal settings – for example, when a sense of collective identity emerges in response to an emergency (Drury *et al.*, 2009) – shows that sharing a social identity with others can contribute to individuals' well-being (Jetten *et al.*, 2012). Again, the underlying process seems to derive from the sense of 'we-ness' which leads people to believe their fellows will help.

We investigated whether attending the Mela impacted on well-being with a survey of a sample of 416 Kalpwasis and 127 comparable others who did not attend the Mela (see Tewari *et al.*, 2012). Both samples were visited in their home villages in December 2010 (before the Mela) and again in March 2011 (after the Mela), and completed questions which gauged their well-being and prevalence of symptoms of ill-health. The two samples were similar in age (one average the Kalpwasis were aged 64, the Controls, 61), gender (57 per cent of the Kalpwasis were female, 50 per cent of the Controls were female) and caste (Kalpwasis: 92 per cent General Caste (GC), 8 per cent fall into the Indian Government classification of Other Backward Caste (OBC); Controls: 86 per cent GC, OBC 14 per cent). In the pre-Mela survey these two samples reported very similar levels of well-being and symptoms of ill-health. However, after the Mela, the two samples differed, with those who had undertaken the pilgrimage reporting better well-being and fewer symptoms of ill-health (Tewari *et al.*, 2012).

Of course, attending the Mela does pose real health threats and it would be folly to ignore these. However, for many, participating in the Mela seems to benefit well-being. The mechanisms involved are likely to include the meditative practices and cognitive schemas associated with religious belief that facilitate coping (James and Wells, 2003). Yet, more social processes are likely to be involved. Research comparing the health of religious and non-religious people suggests that the oft-found better health of the former is bound up with the fact that being religious involves participation in social networks and social group activity (Green and Elliott, 2010; Lim and Putnam, 2010; Diener *et al.*, 2011). Indeed, our analyses of our survey data suggest that, in part at least, the health outcomes associated with participating in the Mela are bound up with the experience of a shared identity with one's fellow Kalpwasis and the more intimate social relations this allows (Khan *et al.*, 2015a).

Discussion

Throughout this chapter we have seen that viewing pilgrimage as exotic and beyond the range of our theories of everyday social life limits our capacity to comprehend the pleasures of participation. Given Le Bon's influence, it is all too easy to see the pleasures of pilgrimage in terms of a loss of identity and a loss of reason. Moreover, following Turner's (1973) characterisation of pilgrimage as liminal to everyday life, it is easy to slip into seeing the pleasures of pilgrimage as reflecting a rather mystical immersion in a sea of common humanity (Turner's *communitas*). Set alongside colourful images of naked bodies and bearded holy men, the product is a picture of pilgrimage as mysterious and mysteriously powerful.

As an alternative, we have argued that there is much to be gained through looking at pilgrimage through the lens of theories developed to address everyday social processes. Specifically, we have considered how the social identity approach to group behaviour offers a framework with which to explore just how and why people in collective events experience such intensely positive feelings. Our starting point is that to understand pilgrims' experiences and behaviour we must address their social-categorical self-definitions. As we have seen, they may not identify with all co-present others: the range and scope of their identification is an empirical issue and contingent upon various social processes. At the Prayag Melas some will identify as a particular sort of pilgrim (as Kalpwasis) rather than other sorts of pilgrim (for example, 'day bathers'). Yet, even amongst those who are all formally group members (that is, the Kalpwasis), we cannot assume that all will experience a sense of shared identity. Perhaps because of disagreements over the nature of certain ritual practices or because of more mundane disagreements over the use of shared water or toilet facilities, some Kalpwasis will not identify with their neighbours. Moreover, even if a Kalpwasi does identify with these others, it is not necessarily the case that these others would reciprocate. To identify with a group is one thing, however to have a sense of shared identity with others it is crucial that others recognise you as a fellow group member. Others' recognition of one's common identity can be exhibited in all manner of ways (for example, the greetings one receives: Hopkins and Greenwood, 2013) and evidence from other collective events (a St Patrick's Day celebration: Pehrson *et al.*, 2013) shows that where participants believe that their sense of shared identification with others is not reciprocated, the experience can be profoundly dispiriting and alienating.

Our key point is that the pleasures of pilgrimage are intelligible. To the degree that one identifies with others and believes this to be reciprocated (such that there is a sense of shared identity), one's experience will be positive. A shared social identity allows an alignment of goals and facilitates the social co-ordination of behaviour such that individuals can better realise the injunctions associated with their pilgrim identity (for quantitative data on this see Khan *et al.*, 2015b). Moreover, a shared social identity provides a resource when the going gets tough which can contribute to one's health and well-being after the event is over (Khan *et al.*, 2015a). Of course, other people's company is not always uplifting, and as Sartre's character Garcin intimates (in *No Exit*), hell can indeed be 'other people'. However, with a sense of shared social identity one's fellows may not only be less hellish but may contribute to one's pilgrimage experience. Indeed, and as we saw in an earlier quote (in which a woman spoke of the pleasures of being with others who also sought to lead a simple spiritual life), others can help one create a little heaven on earth such that 'for one month we see heaven'.

Acknowledgements

The authors thank Professor Steve Reicher, Professor Narayanan Srinivasan, Professor Mark Levine, Dr Gozde Ozakinci and Dr Tushar Singh. This research was made possible by an ESRC research grant 'Collective participation and social identification: A study of the individual, interpersonal and collective dimensions of attendance at the Magh Mela' (RES-062–23–1449).

References

Abubakar, I., Gautret, P., Brunette, G.W., Blumberg, L., Johnson, D., Poumerol, G., Memish, Z.A., Barbeschi, M., and Khan, A.S., 2012. Global perspectives for prevention of infectious diseases associated with mass gatherings. *Lancet Infectious Diseases*, 12: 66–74.

Bhardwaj, S.M., 1973. *Hindu Places of Pilgrimage in India*. Berkeley: University of California Press.

Bleie, T., 2003. Pilgrim tourism in the Central Himalayas. *Mountain Research and Development*, 23: 177–84.

Cassidy, C., Hopkins, N., Levine, M., Pandey, J., Reicher, S. and Singh, P., 2007. Social identity and collective behaviour: some lessons from Indian research at the Magh Mela at Prayag. *Psychological Studies*, 52: 286–92.

Coleman, S., 2002. Do you believe in pilgrimage? *Communitas*, contestation and beyond. *Anthropology Theory*, 2: 355–68.

Coleman, S. and Elsner, J., 1995. *Pilgrimage Past and Present: Sacred Travel and Sacred Space in the World Religions*. London: British Museum Press and Cambridge MA: Harvard University Press.

Collins-Kreiner, N., 2010. The geography of pilgrimage and tourism: transformations and implications for applied geography. *Applied Geography*, 30: 153–64.

Diener, E., Tay, L. and Myers, D., 2011. The religion paradox: if religion makes people happy, why are so many dropping out? *Journal of Personality and Social Psychology*, 101: 1278–90.

Drury, J. and Reicher, S.D., 2000. Collective action and psychological change: The emergence of new social identities. *British Journal of Social Psychology*, 39: 579–604.

Drury, J., Cocking, C. and Reicher, S.D., 2009. The nature of collective resilience: Survivor reactions to the 2005 London bombings. *International Journal of Mass Emergencies and Disasters*, 27: 66–95.

Durkheim, E., 1912 [1995]. The *Elementary Forms of Religious Life*.Translated by K.E. Fields, 1995. New York: Free Press.

Durkheim, E., 1913. Le problème religieux et la dualité de la nature humaine. *Bulletin de la Societé Française de Philosophie XIII*, 13: 63–100.

Eade, J. and Sallnow, M., eds, 1991. *Contesting the Sacred: The Anthropology of Christian Pilgrimage.* London: Routledge.

Ehrenreich, B., 2007. *Dancing in the Streets.* New York: Metropolitan.

Foley, R., 2013. Small health pilgrimages: Place and practice at the holy well. *Culture and Religion:* An Interdisciplinary Journal, 14: 44–62.

Green, M. and Elliott, M., 2010. Religion, health and psychological well-being. *Journal of Religion and Health*, 49: 149–63.

Haslam, S.A. and Reicher, S.D., 2006. Stressing the group: Social identity and the unfolding dynamics of stress. *Journal of Applied Psychology*, 91: 1037–52.

Haslam, S.A., O'Brien, A., Jetten, J., Vormedal, K. and Penna, S., 2005. Taking the strain: Social identity, social support, and the experience of stress. *British Journal of Social Psychology*, 44: 355–70.

Hopkins, N., 2011. Dual identities and their recognition: minority group members' perspectives. *Political Psychology*, 32: 251–70.

Hopkins, N. and Greenwood, R.M., 2013. Hijab, visibility and the performance of identity. *European Journal of Social Psychology*, 43: 438–47.

Hopkins, N. and. Kahani-Hopkins, V., 2004. The antecedents of identification: A rhetorical analysis of British Muslim activists' constructions of community and identity. *British Journal of Social Psychology*, 43: 41–57.

Hopkins, N. and Kahani-Hopkins, V., 2009. Reconceptualising 'extremism' and 'moderation': From categories of analysis to categories of practice in the construction of collective identity. *British Journal of Social Psychology*, 48: 99–113.

Jaffrelot, C., 2009. The Hindu nationalist reinterpretation of pilgrimage in India: the limits of Yatra politics. *Nations and Nationalism*, 15: 1–19.

James, A. and Wells, A., 2003. Religion and mental health: Towards a cognitive-behavioural framework. *British Journal of Health Psychology*, 8: 359–76.

Jetten, J., Haslam, C. and Haslam, S.A., eds, 2012. *The Social Cure: Identity, Health, and Well-being.* London and New York: Psychology Press.

Joyce, C., Stevenson C. and Muldoon O., 2012. Claiming and displaying national identity: Irish Travellers' and students' strategic use of 'banal' and 'hot' national identity in talk. *British Journal of Social Psychology*, 52: 450–68.

Kamitsis, I and Francis, A.J.P., 2013. Spirituality mediates the relationship between engagement with nature and psychological wellbeing. *Journal of Environmental Psychology*, 36: 136–43.

Kawachi, I., Subramanian, S.V. and Kim, D., 2008. *Social Capital and Health.* London and New York: Springer.

Khan, S., Hopkins, N., Reicher, S., Tewari, S., Srinivasan, N. & Stevenson, C. (2015a). Shared identity predicts enhanced health at mass gatherings. *Group Processes and Intergroup Relations.* DOI: 10.1177/1368430214556703

Khan, S., Hopkins, N., Reicher, S., Tewari, S., Srinivasan, N. & Stevenson, C. (2015b). How collective participation impacts social identity: A longitudinal study from India. *Political Psychology.* DOI: 10.1111/pops.12260

Le Bon, G., 1895. *The Crowd: A Study of the Popular Mind.* London: Ernest Benn.

Levine, M., Prosser, A., Evans, D. and Reicher, S., 2005. Identity and emergency intervention: How social group membership and inclusiveness of group boundaries shape helping behaviour. *Personality and Social Psychology Bulletin*, 31: 443–53.

Lim, C. and Putnam, R. D., 2010. Religion, social networks and life satisfaction. *American Sociological Review*, 75: 914–33.

Maclean, K., 2008. *Pilgrimage and Power: the Kumbh Mela in Allahabad, 1765–2001*. New York: Oxford University Press.

Maddrell, A. and della Dora, V., 2013a. Editorial: Spaces of renewal. *Culture and Religion: An interdisciplinary Journal*, 14: 1–7.

Maddrell, A. and della Dora, V., 2013b. Crossing surfaces in search of the Holy: landscape and liminality in contemporary Christian pilgrimage. *Environment and Planning A*, 45: 1105–26.

Mallinson, J., 2013. The making of a mahant: a journey through the Kumbh Mela festival. *The Financial Times* [online], 8 March. Available at http://www.ft.com/cms/s/2/d1e949a0–86bd-11e2-b907–00144feabdc0.html#axzz2qfeuhXhg [Accessed 20 January 2014].

Memish, Z.A., Stephens, G.M., Steffen, R. and Ahmed, Q.A., 2012. Emergence of medicine For mass gatherings: lessons from the Hajj. *Lancet Infectious Diseases*, 12: 56–65.

Messerschmidt, D.A. and Sharma, J., 1981. Hindu pilgrimage in the Nepal Himalayas. *Current Anthrolpology*, 22: 571–2.

Nagar, D. and Pandey, J., 1987. Affect and performance on cognitive task as a function of crowding and noise. *Journal of Applied Social Psychology*, 17: 147–57.

Neville, F. and Reicher, S.D., 2011. The experience of collective participation: Shared identity, relatedness and emotionality. *Contemporary Social Science*, 6: 377–96.

Novelli, D., Drury, J. and Reicher, S. D., 2010. Come together: Two studies concerning the impact of group relations on 'personal space'. *British Journal of Social Psychology*, 49: 223–36.

Olaveson, T., 2001. Collective effervescence and *communitas*: Processual models of ritual and society in Emile Durkheim and Victor Turner. *Dialectical Anthropology*, 26: 89–124.

Pandey, K. Stevenson, C., Shankar, S., Hopkins, N. and Reicher, S.D., 2014. Cold comfort at the Magh Mela: Social identity processes and physical hardship. *British Journal of Social Psychology*, 53(4): 675–90.

Pehrson, S., Stevenson, C., Muldoon, O.T. and Reicher, S.D., 2013. Is everyone Irish on St Patrick's Day? Divergent expectations and experiences of collective self-objectification at a multicultural parade. *British Journal of Social Psychology.* 53(2): 249–64.

Prayag Magh Mela Research Group (PMMRG), 2007. Experiencing the Magh Mela at Prayag: Crowds, categories and social relations. *Psychological Studies*, 52: 311–19.

Reicher. S., 1984. The St Pauls' Riot: An explanation of the limits of crowd action in terms of a social identity model. *European Journal of Social Psychology,* 14: 1–22.

Reicher, S., 1996. 'The Battle of Westminster': Developing the social identity model of crowd behaviour in order to explain the initiation and development of collective conflict. *European Journal of Social Psychology*, 26: 115–34.

Reicher, S.D., 2011. Mass action and mundane reality: an argument for putting crowd analysis at the centre of the social sciences. *Contemporary Social Science*, 6: 433–50.

Renger, D. and Simon, B., 2011. Social recognition as an equal: The role of equality-based respect in group life. *European Journal of Social Psychology*, 41: 501–7.

Sallnow, M.J., 1981. *Communitas* reconsidered: The sociology of Andean pilgrimage. *Man*, 16: 163–82.

Shankar, S., Stevenson, C., Pandey, K., Tewari, S., Hopkins, N. and Reicher, S.D., 2013. A Calming Cacophony: Social identity and the experience of noise at the Magh Mela in Allahabad. *Journal of Environmental Psychology*, 36: 87–95.

Singh, R.P.B., 2011. Politics and pilgrimage in North India: Varanasi between communitas and contestation. *Tourism*, 59: 287–304.

Smith, K.P. and Christakis, N.A., 2008. Social networks and health. *Annual Review of Sociology*, 34: 405–29.

Srinivasan, N., Hopkins, N.P., Reicher, S.D., Khan, S.S., Singh, T. and Levine, M., 2013. Social Meaning of Ambiguous Sounds Influences Retrospective Duration Judgments. *Psychological Science*. 24(6): 1060–62.

Steffen, R., Bouchama, A., Johansson, A., Dvorak, J., Isla, N., Smallwood, C. and Memish, Z.A., 2012. Non-communicable health risks during mass gatherings. *Lancet Infectious Diseases*, 12: 142–9.

Tajfel, H., ed., 1978. *Differentiation Between Social Groups: Studies in the Social Psychology of Intergroup Relations*. London: Academic Press.

Tewari, S. Khan, S., Hopkins, N., Srinivasan, N. and Reicher, S.D, 2012. Participation in mass gatherings can benefit well-being: Longitudinal and control data from a North Indian Hindu pilgrimage event. *PLoS ONE*, 7(10):e47291.doi:10.1371/journal.pone.0047291.

Tully, M., Lannoy, R. and Mahendra, A., 2002. *The Kumbh Mela*. Delhi: Indica.

Turner, J.C., 1982. Toward a cognitive redefinition of the social group. In H. Tajfel, ed., 1982. *Social Identity and Intergroup Behaviour*. Cambridge, England: Cambridge University Press, 15–40.

Turner, J.C., Hogg, M.A., Oakes, P.J., Reicher, S.D. and Wetherell, M.C., 1987. *Rediscovering the Social Group: A self-categorisation Theory.* Oxford: Basil Blackwell.

Turner, V., 1973. The center out there: Pilgrim's goal, *History of Religions*, 12: 191–230.

Tyler, T.R and Blader, S.L., 2000. *Cooperation in Groups: Procedural Justice, Social Identity, and Behavioural Engagement.* New York: Psychology Press.

Wakefield, J.R.H., Hopkins, N., Cockburn, C., Shek, K. M., Muirhead, A., Reicher, S. and van Rijswijk, W., 2011. The impact of adopting ethnic or civic conceptions of national belonging for others' treatment. *Personality and Social Psychology Bulletin*, 37: 1599–610.

Walton, G.M. and Cohen, G.L., 2011. A brief social–belonging intervention improves academic and health outcomes of minority students. *Science*, 331: 1447–51.

Chapter 3

Feelings of the Sacred and their Occurrence in Journeys

David Crouch

Introduction

In this chapter I consider the occurrence of feelings of spirituality and the sacred, not through the popular and academic understandings of sacred journeys, including the often exemplar pilgrimage, but through a reasoning that is open in terms of immanence beyond or outside of that founded in institutions and ritual practices of formal religion. There are overlaps, however, as well as distinctions here, to which I will return. The character and emergence of immanence emerges through the discussions of the chapter. In this particular emphasis, several distinctive threads are treated here in a relational manner. These include multi-disciplinary insights concerning performativity, and the particular consideration of Deleuze and Guattari (1988 [2004]) on the occurrence of becoming and wonder (Dewsbury and Cloke, 2009). Similarly, journeys are considered metaphorically, though not without some possibility of relation with traversing objective linear space.

The chapter has a particular emphasis upon practice and process, the *doing* of such journeys. Discussion of the relation between performance and space draws fresh considerations into meanings surrounding 'sacred' and of 'journeying' and journeys. Performativity is distinguished by an openness to possible consequences. The reconfiguring, or reconstitutive potential of performance is increasingly cited in terms of performativity; as modulating life and discovering the new, the unexpected, in ways that may reconfigure the self and our relations beyond self. Such an approach enables an interrogation of practices, and performativity thus emphasises post-hoc yet pre-cognitive moments and situations in that the acts deemed as performative are those in and through which something unexpected may be brought about, an affect. This 'brought about' can be a rearrangement of existing feelings, embodied insights and relations. The thread of space and how we may *produce* space is attended as a potentially important affect in the potentiality of feeling immanence.

On Journeys

The emergence of such significance in feeling and feelings occurs in a process of performativity that can, in terms of space, be considered as flirting (Crouch,

2010). We 'flirt with space', or place. Flirting with space implies an openness, uncertainty, fluidity and contingency, following Doreen Massey in close-up individual and shared practice and feeling (2005). I have suggested a character of flirting in terms of energies of and between things, and the apparent human need to hold on to some value or meaning of space. Lives, energies in the widest sense, and time are however not fixed, but our actions and feelings can become contexts for later experience. Of course, other cultural contexts influence but do not determine. We flirt (with) space in journeys of our lives, in varying trajectories of time and in the movement or vitality of things; our feeling and intensity. Thus space is lived, touchable, fleshy, also other than merely representational; it shifts and changes with the vitality of life, opens into reassessment, cracking time into relational pockets of significance and so on. Time is mobility in its qualitative multiplicity of open possibility. In our flirting with space, space-times emerge in mobility as the vitality of being alive.

Journeys are coloured by their commingling relationality with space and vice versa. Space and journeys commingle as felt, imagined and projected. Journeying in this sense is material and metaphorical; journeys are in the liveliness of energies that we may discover happening. We discover also a desire for openness, becoming; of 'going further', forward or elsewhere. This going may be realised for some with the help of a material journey of objective distance, or can happen within ourselves. Journeys happen in various trajectories of space-time and its feeling. Sacred journeys of pilgrimage may be slow, endured, uneven, complex and of multiple feeling, affect and response. In pilgrimage, intentional movements may be crucial; a pursuit of particular programmed attainment of recovery or rediscovery, however much they may give opportunity for individual and shared becoming (Maddrell and della Dora, 2013). Moreover there can be numerous things going on *en route*, as the stories of the fourteenth century Canterbury Tales by the English chronicler and author Geoffrey Chaucer testify with some humour. Of course, programmed pilgrimages may not work, they may disappoint. The multiplicities of things that happen in our journeys and their performative moments can be accompanied by being implicitly aware of things and registered in feelings, beyond and other than the familiar; or the familiar can become heightened in awareness, feeling and significance. Changes and different realisations in life frequently happen slowly, arguably with no less possibility for affect. Things happen in low intensity and calmly, even in apparent stillness, that is not to be confused with emptiness (Cocker, 2009). Potential and pace may be affected by the mix of social circumstances and cultural capital. A possibility of variable space and pace opens up a number of threads relating to journeys. Some of these are more redolent with change, others familiarly associated with relative steadiness, and frequently these registers merge, relate over time.

Journeys can be felt to have long gestation but also instantaneous presence; they can occur, for example, in moments of tensions and adjustment in life. Life is, to say the obvious, complex. It has many different journeys simultaneously interwoven. These happen in slowness and acceleration, in the enfolding of different moments and times. My concern here is to provide insight to how these events happen, may become significant, changing or refiguring significances that are felt about things, relations, spaces of life and in relation with a wider, more popular abstract character that we express in terms such as global adjustments, mediated popular culture, directed ritualistic practices and institutional prescription, including their own broader politics.

Moreover change articulates and may commingle with continuity, as nets of forces both complex and vital, rather than in nests of fixity and mobility. The expressions of journeys happen in uttered feelings, words of shared stories, touch and the character of body-movement and other gestures, not of representation but of expression though both communicable and incommunicable; laughter and tears: dynamic, human and emotional processes.

Such vitality makes space for openness and possibility, of becoming. 'Becoming' is explicitly related to the countless possibilities of things, of life, emerging or erupting from what Deleuze and Guattari (1988 [2004]) called an 'immanent surface' of energy. Our pasts, as half-held memories and presents and futures of desire merge. Wonder, often associated with immanence and spirituality, can occur in discovering something other, beyond ourselves and our relations with the materialities of the world, but of ourselves, new or rearranged, in new relations with others of all kinds. Life often works in the tensions and tugs of 'holding on' and 'going further' (Crouch, 2003; Crouch, 2010). These flows are not in crude opposition but multiply emergent and relational. It would seem that we bear traces and desires, at least, of continuity amidst forces, willed, outwardly influenced and also unaware that hang with us and with and within which we become. We may find our journeys irksome or in longing. Alongside and flowing awkwardly amongst these forces we sustain a desire for difference, adventure, things new. This multiple 'tug' is germane of the journeys of living and the fleshy ways in which our spaces emerge and flow. Amidst life's journeys and journeying becoming happens, in stationary reflection that can be also a journey, as Cocker discusses (Cocker, 2009).

The dynamic of becoming happens in vital energies between different forms of intensity and extensivity, low level and hyper-intensity. Different kinds of energy-intensity amongst moments of our encounters with the world contain the potential of different kinds of significance in how we feel, returning to familiar traits and opening to new ones. The space tracked may summarily disappear un-noted but that rootedness may however return in the cracks; and can emerge as powerful and significant: journeys grasp together and mix feelings, memory and the character of practice and performance. Although journeys can be delimited and trapped, this entrapment is only partial and temporal. Time becomes revised and revisited and moulded as our lives journey. The affects that this vitality and complexity of time have can profoundly influence. Many individuals find facility to spirituality in relation to prescribed rituals and journeys that may be in walked across tracts of land, or sat in meditation.

Spaces of Immanence in the Notion of Flirting

How we arrive somewhere, how we feel about where we are, is not scripted but we become participants in its making and meaning. To engage this in what happens and how it feels we may consider the notion of flirting. Flirting opens to energies, Deleuze's immanent surface of infinite possibility. Space is not prefigured or unchangeable, nor is the feeling of space. Amidst this complexity we flirt with space and with time, or its commingled character as space-time. Amidst the dynamic of flirting the character and feeling of space-time can change. The unexpected in the familiar; the significance rather than superficiality of both mundane and brief moments *en route* somewhere else, in ourselves and our relational lives. Ingold suggests something similar when he writes comparing wandering with ideas:

> Each time you revisit the (space) it is a little different, enriched by the memories and experience of your previous visit. Leading others along the same pathways, you may share the idea with them, though again, as each brings along the peculiarities of their own experience, it will not be quite the same for one individual as anyone else. (Ingold, 2007: 9)

Yet there is a wider embrace: space is occurent within and between us, relationally and also with other life and extant materiality, and happens amongst a multiplicity of our own and others' space-times.

Journeys can be more than individual and private; they are frequently inter-subjective in absence and presence. They occur in and amongst instants and moments but act relationally with time. Our pasts are mutually eloped, unevenly and awkwardly enfolded in this mass of convolutions, challenged and affirmed. Moments in journeys are not isolated, but prompt and are prompted by other loops and re-loops, temporary suspensions, threads of that commingling of space and time as space-time of life. Memory and feeling are not simply 'placed' or positioned in time in a linear 'order-ing' of being, but tumble amongst others, or exist in a net with others, open to being re-grasped anew in other moments. Memory can be rediscovered. In and out of these flows are inflected feelings across a range of being, dwelling and becoming. In journeys our feelings about ourselves and our relationships in the world are negotiated but also happen to us. Immanence can emerge from this creativity.

Performativities, Space and the Sacred

In this section I activate these theoretical reflections in an examination of a repositioning of conceptual reasoning around sacred journeys relationally with ideas of space and its making-significance. Not least I suggest a shift from essentialised space-related actions and events and their affects to matters of living and becoming: of opening to potentialities. I consider approaches to secular and sacred, and reflect upon the heavy yet uncertain weight that these terms or notions carry. The idea of loosening the categories of conceptualising sacred journeys works alongside two core presumptions. One is the habitual emphasis of the sacred in terms of distinctive institution-led or related spiritualities, from which or in relation with which ideas, attitudes and feelings of the sacred may emerge. The other concerns the understanding of journeys as containing at least some geographical-measurable distance to cross and often to endure, thus becoming reified in sacred character. Arguably, space and its potentiality for immanence are much more than this.

In Doreen Massey's geography space is a participatory and dynamic energy: 'the coming together of the previously unrelated, a constellation of processes rather than a thing. This is place as open and internally multiple ... not intrinsically coherent' (Massey, 2005: 141). Space is more than the contextual co-ordinates of the social, economic and political; more than the materials and their physical and metaphorical assemblage, of building material, vegetation and rock. Space is increasingly recognised to be always contingently related in flows, energies and the liveliness of things; therefore always 'in construction', rather than fixed and certain, let alone static (Massey, 2005). What space 'is' and how it occurs is crucially rendered unstable and shifting; matter and relations in process. It may be felt to be constant, consistent and uninterrupted, but that feeling is subjective and contingent.

The energy and vitality of space is articulated in the work of Deleuze and Guattari that has helped unravel and unwind familiar philosophies of the vitality of things; the multiplicities of influences and the way they work; and in a world of much more than the result of human construction. They offer a means to rethink the dynamics of space (Deleuze and Guattari, 1988 [2004]). Our feelings work amongst the intensive capacity to affects and to be affected. Their term spacing introduces a fresh way of conceptualising the process-dynamics of the unstable relationality of space/life. Spacing occurs in the gaps of energies amongst and between things, in the cracks of everyday life as much as in the prescribed ritualised practice; in their commingling.

Thus 'space', felt, emerges in a panoply of affects: of external materiality and other-than human life (animals, plants and so on), our self and our relations with others, with mediated and other contexts commingling, but not determining. Our own feeling and grasp, alongside our practical ontology, become also part of the contexts of future experience, moments and feelings. And we affect what is around us, also relationally amongst the affects of the non-human (Lorimer, 2006). Moreover, a feeling of wonder and heightened, or simply different, intensities and an awareness that familiarly accompanies feelings of the sacred, can involve emotions from life.

Recent debate on affect offers a further dimension to the character of the emergence of immanence:

> ... you can think of affect in the broadest sense as what remains of the potential after each or everything a body says or does – as a perpetual bodily reminder.

> ... a reserve of every actual production of meaning in language or in any
> performance ... vaguely but directly experienced [felt] as something more, a
> more to come, a life over-spilling as it gathers itself to move on. (Massumi and
> Zournazi, 2002: 213)

A feeling of sacred, or spirituality, or simply of immanence may therefore take
us into, or open up, or make felt, other parts of ourselves, our being. Sacred does
not require an Other (that is, as a supreme 'it', another entity), but is the perhaps
momentary discovery or rediscovery of other dimensions, registers, intensities,
alignments of feeling in and of our own being, and becoming. Contained in such
emergence is the potentiality of being alive. Such a position aligns aside notions
of dogmatic realism, objective and ordered and coded existence in relation to a
cosmos or some such, existing independently of and non-relationally with our own
materiality. Feelings of wonder reasoned thus are not prefigured, though in the
familiarity of assurance and holding on, we may feel that we want to hitch ourselves
to the pre-figured. Arriving into a feeling of becoming may be slight or very present;
may feel disorienting or reassuring; or full of tensions between the two.

In Michael Hall's discussion of such an arena of experience he draws
perceptively upon the work of Don Cupitt, an English theologian who decades ago
turned to what he calls Christian religious humanism, defining god as ' ... the sum
of our values, representing to us their ideal unity, their claims upon us and their
creative power' (Cupitt, 2008, quoted in Hall 2006: 67). Hall notes that people
strive to find ultimate harmony and unity of moral and spiritual values, yet that
are impossibly distant from what we can achieve, though transcendental moments
may be felt. Hall also mentions Iyer's assertion that ' ... a pilgrim does not have
to be moving towards something holy ... so much as towards whatever resides in
the deepest part of him (sic.)' (Hall, 2006: 74).

What Hall expresses from Cupitt is the character of becoming. Becoming may
arise unexpectedly or expectedly in joining a particular route towards meaning, or
feeling, of a particular kind. Thus feeling and meaning are not coincidental, but
may emerge at the same moment, or not. Reflection might position a *knowing* in
alignment with the feeling. However that potentiality may be limited in certain
routines of religious conformity; routes to feeling already prescribed into which
we organise a formal knowing, that may deny or ignore the potentialities of a
broader wonder or heightened awareness from the moment itself. However, as
Maddrell and della Dora, like Cupitt and also Coleman and Eade have considered,
there is frequently an openness in conformative practice too (Maddrell and della
Dora, 2013; Coleman and Eade, 2004; Cupitt, 2008).

As a Quaker, I am perhaps attuned to paths and tracks that may open to us in
our everyday apparently mundane living, perhaps more so than more formalised
religions which may prioritise structured procedure and pathways. Quakers have
only laity, negligible doctrine, welcome all faiths and include humanist Quakers. In
thinking through such potentialities the notion of feelings appeals, it both contributes
to compose the cultural and can activate and participate in the feeling of immanence.

My particular route of consideration relates to our relationship in the character of space. Rather than consider the literal application of mobility, as Coleman and Eade tend towards (2004) I consider the quality of space and its emergence through the notion of flirting introduced earlier in this chapter. Into this swirl of occurrences, affects and embodied performativities, feelings occur, with their resonance and affects of senses and the materiality of the world in which we are.

The anthropologist Kathleen Stewart draws similar attention to the swirl of related events, none given primacy, in the ways that we make sense, fleetingly, incompletely, in flows, but significantly relationally, in her term 'worlding':

> We are left with acts and gestures and immanent possibilities ... it tracks the pulses of things as they cross each other, come together, fragment and recombine in some new surge. It tries to cull attention to the affects that arise in the course of the perfectly ordinary life as the promise, or threat, that something is happening – something capable of impact. ... they take us to the surge of immanence itself. (Stewart, 2005: 1029)

A few months before writing this piece I wrote a paper concerning belonging and being at home in the world. Perhaps paradoxically, as I came to complete the current piece, I discovered how close, how merging, sacred journeys can be to the energies, happenings and feeling that nudge our belonging or disorientation. Journeys can happen between space-times. Our memory of other space-times can interrupt any linear flow we may experience. Journeys become, emerge as our own life experience, in practices and performativities. Our imaginative spaces are embodied relationally.

Brian Massumi approaches the hopes and potential for freedom amidst oppression and control, writing concerning the practice of everyday life, ethical practices, belief in the world that makes us care for our belonging as well as change, belonging in change; the passing awareness of being at a threshold' (Massumi and Zournazi, 2002). As anthropologist Kathleen Stewart puts it: 'Ordinary affects are the varied, surging capacities to affect and to be affected that give everyday life the quality of a continual motion of relations, scenes, contingencies and emergencies. They're the things that happen' (Stewart, 2007: 1).

Frequently the writing of Deleuze and Guattari (1988 [2004]) on immanence, 'energy' and 'forces' feels suggestive of a sense of spirituality, however vague or implicit, or absent. Yet there persists a feeling of resonance with what I can only call my 'grasp' on something like spirituality. Dewsbury and Cloke have considered the notion of 'spirituality' in terms of Deleuze and others (Dewsbury and Cloke, 2009). Their perspective focuses explicitly on Christianity. My interest being unbound by tradition and any institutional prefiguring of things sacred, the present discussion is inflected more by a notion of an open 'enchantment' that Bennett considers a kind of sensation. In this there is feeling beyond the materially-constituted and beyond the socially-culturally constructed. Bennett's notion of enchantment is ' ... to be struck and shaken by the extraordinary that lives amid the familiar and everyday ... enchantment entails a state of wonder ... a momentarily immobilising encounter; it is to be transfixed, spellbound' (Bennett, 2001: 4). Of course, again as Maddrell and della Dora and also Coleman and Eade have said, such a feeling can be endemic to much religious sacred journeying (Maddrell and della Dora, 2013; Coleman and Eade, 2004). In this chapter my emphasis continues upon other ways and occurrence of such potential.

'A Kind of Immanence'

Flirting with space enables a breaking through continuing meta-thinking to attend to people's living in the world, partly as quiet or gentle politics; partly as profoundly significant influence and affect on the meaning and value of things and the way things happen. Across the several voices expressing feeling in flirting with space in my research on the so-called everyday, or ordinary, there seems to be an element of these: in 'humble' (*sic*) caravanning and doing of community gardening 'my little bit kingdom' (Crouch, 2001; 2003; 2010); in the moment of achieved tension in making artwork and in its circulation of participative encounter (Crouch and Toogood, 1999; Crouch, 2010). I find this in Owain Jones's feeling of belonging, who insists on atheism, with which there need be no contradiction in terms of my loose notion of 'enchantment' (Jones, 2005). Moments, events and encounters can be felt deeply: so what may we mean by that? Sacred journeys, feelings of the sacred, are not limited to the religious.

For example, in a sensitive essay Owain Jones recalling his 'home' expresses autobiographically emotional significances of belonging (2005). He is concerned particularly with the merging of memory and contemporary, happening, feelings of belonging: drawing disoriented alignments in recall, connecting and feeling loss. He says 'Memories mobilise, a landscape within me comes alive, yet into something fresh. I change' (Jones, 2005: 217). Jones' belonging and identity are not fixed in a particular space-time, but draw upon it.

Fluid time persists as an awkward, chaotic thing; an awkwardness of belonging in moments of presence that resonate the way in which sociologist Ann Game conceptualised the dynamic character of belonging. Belonging is often conceptualised in the nostalgic characterisation of the past. In contrast Game (1991) argues for a belonging that is experienced in our everyday living yet not divorced from memory and which emerges out of feeling like a child:

> Moments when we feel wide-eyed, wide open, in love with the world. Running into the waves, the salt-smell spray in my face, or feeling the sand between my toes ... these are moments of feeling 'this is right', 'now I have found what I have always been looking for, what I have always known'. I get that 'coming home' feeling ... that might best be described as a sense of belonging. (Game 1991: 227)

Belonging emerges through the duration of life and across its journeys, both momentary and over a long trajectory; they produce flows of time that can be detonated into significance.

Norwegian geographer Inger Birkeland narrates an unfamiliar wonder of being that transforms her from feelings of detachment from the familiar and entwines her in something much more significant, a seemingly 'remote' sense of belonging, visiting the Arctic Circle at midsummer in Scandinavia:

> In the evening I was waiting for the deep red midnight sun. I was alone but didn't feel lonely. We were many who shared the act of waiting for the midnight sun. ... Even if we were strangers to each other, there was a mutual feeling of waiting for the midnight sun ... as more and more visitors arrived at the cliffs, I felt like I was walking in a multicultural, multicoloured city ... The words uttered were the uncomplicated, the kind of words that sound trivial outside the there and then. But they were not trivial, rather they represented another way of creating meaning out of the meaningless, Order out of Chaos, light out of darkness. (Birkeland, 1999: 17)

The heightened awareness is at once both individual and collective, shared. Inger explores her experience through the writing of Kristeva, as engaging the poetic character of our experience of being contained in the world as in the womb: 'A womb-like state where all needs are (felt to be) capable of being fulfilled' (Birkeland, 1999: 27). Her experience felt shared yet semi-attached. The iconography of the 'Northern Lights' provided an aspect of anticipation, but the moment became her own and 'immediately took us beyond an immediate feeling of fun and enjoyment to one of wonder' (Birkeland, 1999: 27). She finds the journey complex, away from home but belonging, somewhere unawares became intimate. In the moment of her performativity creativity erupts as a change of register, not in habituation or ritual.

The site-specific sculptor and installation artist John Newling considers moments of 'arrival' at a new site for his work: 'when I cross the threshold of a space that might house a project, I am aware of a transition of thinking. The relationship between the threshold and what the space could hold is undermined, open and ambiguous' (Newling, 2007: 38). He acknowledges the contexts, pasts and other presents that a space can have, and calls this collision of feelings and awareness, openness, disorientation, the character of the tacit agreements individuals may have had and still have with the space are dislodged in his mind and in the project that may follow. 'What was familiar becomes uncertain and then begins to reconstruct its familiarity into new possibilities' (Newling, 2007: 39). He calls this disorientation 'a change in the relationship between time, place and person' (Newling, 2007: 41). These artworks operate in the edges, in the gaps of spacing, the numerous spacings amongst things and individuals' lives and feelings. In this work there is the character of liminality: full of potentiality. Indeed, taking the notion of flirting with space further, all sites or spaces bear the potential of liminality (Crouch, 2012).

Pertinently, an example of Newling's work is Chatham Vines. Growing vines hydroponically in a redundant church in southeast London, he explains:

Intoxication

> The material and symbolic journey of the grapes reveals a conceptual aspect of the project. Organising inside an abandoned church, the grapes travel through the chemical transformations of the wine making process. The grapes are then transformed again, symbolically, within the service of the Eucharist. More wine will be consumed in a secular environment. The grapes enter our blood bringing the possibility of intoxication'. (Newling, 2006: 63)

The character of enchantment (Deleuze and Guattari 1988 [2004]), whose work ironically feels to be of the spiritual, is pursued directly by Dewsbury and Cloke. They provide a formative way of thinking and feeling about spirituality: 'the spiritual' as 'that part of the virtual in which faith forms a significant part of the move beyond rationality and of the possibility of other-worldly disposition' (Dewsbury and Cloke, 2009: 696). As Kong asserts, 'place' or space as sacralised does not rely upon institutional framing, figuring or permission (2001). Deleuze and Guattari consider 'the spiritual as something constitutive of everyday life; cutting at that space between absence and presence, and manifesting itself at the immediate, and therefore non-metaphysical, level of the body' (Dewsbury and Cloke, 2009: 697). They consider their reflections in relation to 'landscape'. The notion of the poetic may not be equivalent to spirituality. The immanent surface and the 'unaccounted for' in their discussion come close to the notion of energies. John Newling found a similar character of the sacred space in disorientation, again coming back to the relation between belonging in the world and finding immanence (2007).

A spiritual seems to emerge in Deleuze: 'the zone where forms become indiscernible and normal orientation is displaced as understanding comes to be arrested more upon sensible intuition' (Dewsbury and Cloke, 2009: 705). Resembling the notion, energy, dynamic and character of flirting with space, there is a spirituality that is felt, ironically earthy, fleshily disposed thing drawn through living. Deleuze asserted that 'this spirituality is the body without organs' (2003: 46–7).

Metcalfe and Game describe a moment apparently ordinary:

> Of course, I could not see my children like this without seeing the world afresh. When my perspective was lowered by my fear of a tantrum, I had not noticed the glorious open blueness of the sky or the vital greenness of the street trees. This was the first sunny morning after days of rain, and the world was clean and full of promise. As Max and Leo and I walked to school, hand in hand, I could feel the world smiling at us, with us, through us. (Metcalfe and Game, 2008: 21)

This sense of delight in the world, of opening in becoming towards feeling beyond the material, though affected in part by the material, has the character of spirituality. As Maddrell and della Dora have considered with particular regard to contemporary Christian pilgrimage (2013), these expressions are embodied, yet do not necessarily carry ritual, and occur in the liminality of life and its relations in the occurrence of space.

Arising Reflections

In this chapter I have sought to open the frame of what may be familiarly marked as 'sacred' and sourced in 'journeys'. Journeys and the possibility of the experience or feeling of immanence are not, I argue, confined to institutional labelling, direction, organisation and quest. There is, moreover, no necessary separation between the institutional and the individual/collective. Through these diverse considerations I have sought to draw together some key insights that make up an understanding of process, practice and emergence of more-than 'where we feel we are' in life; changes of registers, intensity, rhythm and depth of feeling. These can each be mixed, commingled or felt of intense clarity. Feeling is close to the moment of experience of the sacred. Perhaps through these lines of thought we can discover different ways of thinking and explaining matters of space, journeys, 'the sacred' and our relations in the world.

References

Bennett, J., 2001. *The Enchantment of Modern Life: Attachments, Crossing and Ethics*. Princetown, NJ: Princetown University Press.

Birkeland, I., 1999. The mytho-poetic in northern travel. In D. Crouch, ed., 1999. *Leisure/tourism Geographies*. London: Routledge. 17–33.

Cocker, E., 2009. Stillness. *Journal Media-culture*, 12(1): 2–13.

Coleman, S. and Eade, J., eds, 2004. *Reframing Pilgrimage: Cultures in Motion*. London: Routledge.

Crouch, D., 2001. Spatialities and the feeling of doing. *Social and Cultural Geographies*, 2(1): 61–75.

Crouch, D., 2003. Spacing, performance and becoming: the tangle of the mundane. *Environment and Planning A*, 35: 1945–60.

Crouch, D., 2010. *Flirting with Space: Journeys and Creativity*. Farnham: Ashgate.

Crouch, D., 2012. Afterword. In L. Roberts and H. Andrews, eds, 2012. *Liminal Landscapes: Travel, Experience and Spaces In-between*. London: Routledge, 234–41

Crouch, D. and Toogood, M., 1999. Everyday abstraction: geographical knowledge in the art of Peter Lanyon. *Ecumene*, 6 (1): 72–89.

Cupitt, D., 2008. *Above Us Only Sky*. Santa Rosa, CA: Polebridge Press.

Deleuze, G. and Guattari, F., 1988 [2004]. *A Thousand Plateaus*. London: Athlone Press.

Deleuze G., 2003. *Francis Bacon: The Logic of Sensation*. London: Continuum.

Dewsbury, J-D. and Cloke, P., 2009. Spiritual landscapes: existence, performance and immanence. *Social and Cultural Geography*, 10(6): 1464–9365.

Game, A., 1991. *Undoing the Social: Towards a Deconstructive Sociology.* Buckingham: Open University Press.

Hall, C., 2006. Travel and journeying in the Sea of Faith: perspectives from religious Humanism. In D.J. Timothy and D. Olsen, eds, 2006. *Tourism, Religion and Spiritual Journeys*. London: Routledge, 64–77.

Ingold, T., 2007. *Lines: A Brief History*. London: Routledge.

Jones, O., 2005. An emotional ecology of memory, self and landscape. In J. Davidson, L. Bondi and M. Smith, eds, 2005. *Emotional Geographies*. London: Ashgate, 205–18.

Kong, L., 2001. Mapping 'new' geographies of religion: politics and poetics in modernity. *Progress in Human Geography*, 25(2), 11–33.

Lorimer, H., 2006. Herding memories of humans and animals. *Environment and Planning D: Society and Space*, 24: 497–518.

Maddrell, A. and della Dora, V., 2013. Crossing surfaces in search of the holy: Landscape and liminality in contemporary Christian pilgrimage. *Environment and Planning A*, 45(5). 1105–26.

Massumi B., in conversation with Zournazi M., 2002. Navigating movements. In M. Zournazi, ed., 2002. *Hope: New Philosophies for Change*. Annandale, NSW: Pluto Press, 210–42.

Massey, D., 2005. *For Space*. London: Sage.

Metcalfe, A. and Game A., 2008. Potential space and love. *Emotion, Space and Society*, 1(1): 18–21.

Newling, J., 2005. *An Essential Disorientation*. Poland: SIRP.

Newling, J., 2006. *Chatham Vines*. London: Artoffice.

Stewart, K., 2005. Cultural Poesis: the generativity of emergent things. In N. Denzin and Y. Lincoln, eds, 2005. *Handbook of Qualitative Research*, 1015–30

Stewart, K, 2007. *Ordinary Affects.* Durham NC: Duke UP.

Chapter 4
Mobilities of Magick

James Thurgill

Introduction

To welcome in the Pagan New Year festival of Samhain 2011, 55 people gathered together in preparation of an evening ritual at a National Trust car park in the Middlesex village of Wraysbury, UK. As the surroundings grew darker, the group made their way by candlelight through the conservation land until they arrived at their final destination, the Ankerwycke Yew (National Trust, 2010). A tree of colossal scale in regards to both its size and age, the Ankerwycke Yew acts as a sacred place for ritual goers, attracting worshippers throughout the year. The tree itself functions as the focal point of the ritual, it is where believers hold that the sacred becomes manifest. However, there is significant ritualistic importance placed upon the journey between the meeting point and the Yew itself which should not be overlooked, a commonality that exists between the movement of actants in and around Neo-Pagan ritual and early forms of Celtic-Christian pilgrimage, a removal of the normative demarcations of the divine in favour of 'sacralising the quotidian' (Maddrell and della Dora, 2013: 1109). There were a number of stoppages that occurred in the space between congregation and ritual during the procession; magical practices, performances and invocations took place along the way and I will use these examples to form an argument for the importance of mobility in magickal practice. It will be shown that these moments of delay when making the sacred journey are not superfluous to the ritual performance, rather they are integral to it, heightening a sense of the sacred and galvanising the magickal ritual space in a pilgrimage of sorts.

Through an analysis of both the ritual and the preceding journey towards the Yew, I want to explore the vital role mobility plays in the processes of magick and the formation of an enchanted place. I will begin by providing a descriptive auto-ethnographic account of the ritual that took place on Halloween 2011, showing how the application of mobility to the magickal practices conducted at the Ankerwycke Yew act to both augment a sense of its sacrality and engender a process of place-making. I position myself fully within the ritual, reflecting upon how the practices I encountered affected me, both physically and emotionally, elucidating the manner in which site and sacrality impacted upon my sensory engagement with the space. I will then discuss the conceptual theory of both magick and enchantment in regards to place and mobility, moving on to show how these processes are used in the fostering and conservation of magickal ecologies; spatialities that allow for the extraordinary to occur.

The focus of this chapter is not so much to provide a space for the further analysis of mobility itself; rather it is to form a discourse around enchantment, magick and place that acknowledges the role of movement as inseparable from the performance of the sacred in nature-based worship and practice. Places themselves are all about connections (Massey, 1994); sacred places are no different, they too require connections, to the people who use them and to the landscape they occupy. This connectivity is itself underpinned by mobility; a purposeful movement to and from the site as well as the circulation of materials, myth, worship and ritualistic practices within and around the sacred space that promote and extend spiritual importance through a process of reiterated performances that seek to develop the spiritual or imaginal landscape (Blain and Wallis, 2004). Mobility proffers a deliberate method for opposing the world as inanimate, static (Wylie, 2007) and instead sees places as existing within a continuous state of flux. Reflecting the ways in which we experience the world through a set of embodied practices, mobility emphasises the production of spatial narratives (Cresswell and Merriman, 2011). Maddrell and della Dora (2013) examine the links between movement, place and landscape through 'surface framework', a response to both the vertical and horizontal trajectories of pilgrimage which allows for a deepened analysis of sacred mobility via the process of 'embodied travel, ritual, perceptions of the visual, material, affective, and the numinous' (Maddrell and della Dora, 2013: 1105). Pilgrimages act as 'kinetic rituals', working to charge space with a sense of sacrality and further determining their function as places of spiritual importance (Coleman and Eade, 2004).

The movement that occurs between sacred sites forms part of a process that sees the ongoing production of the sacred space; it is not just place and ritual but also movement that works to define the sacrality. Cresswell claims that 'places are never complete, finished or bounded but are always becoming – in process' (Cresswell, 2002: 20) and although this transience might well appear to undermine the sense of a fixed sacred in many domains, we shall see how this strengthens the affective properties in cases of sites with an ascribed magickal sacrality. In such instances the idea of a placio-spatial becoming can be seen to support the idea of a vital, animate sacred in which to engage with; a landscape imbued with a sense of super-natural agency. The presence of a vital mobile place that 'gathers things, thoughts and memories' (Escobar, 2001: 143), not only positions mobility as a key facet in the theorising of a site, but additionally places it within a spiritual framework that at once aims to connect with the observer (so as to provide an experience of place) whilst simultaneously 'gathering' them into becoming part of the place, part of the place's materiality. This is the case with the Ankerwycke Yew, a place that requires mobility in its ritualistic worship, its consequent spatial performance and entering into the magickal.

The Ankerwycke Yew is located within a small wooded area alongside the River Thames approximately one mile south of the Middlesex village of Wraysbury, a

Figure 4.1   The Ankerwycke Yew, Wraysbury, UK
Source: Author's own photograph

sleepy semi-rural community situated 18 miles west of London, UK. As one of only 50 trees listed as heritage by the National Trust, the Ankerwycke Yew is one of the oldest trees in Britain, believed to be between 1,400 and 3,000 years old (Bevan-Jones, 2002). The Yew sits within the former grounds of Ankerwycke Priory. The site is bordered to both the east and south by the River Thames, which meanders through this small area of land between the counties of Surrey, Berkshire and Middlesex. The Yew is enclosed by tree lined ditches on all sides. It is pertinent that this spectacular monolith sits isolated from local communities, the dense woodland and boundary waterways act as a moat for the site that has remained in situ for millennia. Furthermore, the bordering of the area by water and woodland means that the site is a quiet, secluded place which aids in its use within ritualistic practice.

On visiting the site, what immediately becomes clear is the abnormal scale of the tree, it appears out of place amongst the otherwise modestly sized trees of the British woodland. Recorded as being 27ft 8 inches in girth in 1822 (Bevan-Jones, 2002: 58) the Yew has now achieved a circumference of over 31ft. around (nationaltrust.org). There are various historical narratives that tie into the tree's location. Reference is made to the Yew in Charles Knight's *Old England: A Pictorial Museum* (1845) where Knight claims that the ancient tree was a meeting place of Henry VIII and Anne Boleyn. Delving further still into the area's past; the Magna Carta of 1215, originally thought to have been sealed at the other side of the Thames in Runnymede, is now considered to have taken place at Ankerwycke. Indeed both the positioning of the site off of Magna Carta Lane and its close proximity to the Wraysbury parish of Magna Carta Island, help to feed this notion. Runnymede, situated on the southern banks of the Thames just a few hundred metres from the tree itself, is derived from the Saxon, meaning 'meadow of runes' (Carr-Gomm and Heygate, 2010), deepening the Yew's place within an ecology of magick.

The tree is scarred, split, entangled with itself through a process of rupture and re-growth; its bark, bough and branches tell the tales of a long and arduous existence. The Yew is by no mistake a tree, this much is obvious insofar as can be deduced from the structure of its matter and resemblance or similitude to all other trees. And yet, such a tree feels different, its character unlike those of trees that are surrounding it. The twisted, hollowed out trunk serves to stricken this sense of enchantment further still, an 'empty vessel' that both hosts and amplifies the contradicting nature of the tree as both animate and static, vernacular yet sacred (Eade and Sallnow, 1991). In viewing the northern side of the trunk, a certain anthropomorphism occurs; there appears to be faces, bodies and limbs embedded and entangled within the tree itself. The tree seems almost human and yet unfathomably inhuman at the same time.

Studying the tree closer one notices the vast array of ribbons, scraps of paper, crystals and polished stones that appear to act as offerings to the Yew. Outside of its normative historical accounts, Ankerwycke is a site for Neo-Pagan ritual and ceremony, a tradition that I will explicate later in the chapter. A large broken piece of branch lies to the side of the tree, mounted upon it are the remains of candles, incense and dead flowers. Evidence of ritual, perhaps of magick, objects that signify the tree's significance in Pagan practices – a phenomenon indicative of the ritual activity that has been documented as taking place across sites of Pagan significance in the United Kingdom (Wallis and Blain, 2003). Trees are recognised as symbolic for the Neo-Pagan; they sidle temporal boundaries and offer a stark example of the resilience and longevity of nature, moreover of life. With its beginnings beneath the surface of the earth, its roots anchoring the trunk to the underworld, the body of the tree rises from the ground, claiming dominion by its size alone, *making a place for itself.* The branches of the canopy reach towards the Heavens, extending to the sunlight, the catalyst of all life. For the Neo-Pagan, perhaps nothing embodies spiritual transcendence more than the tree. Pagan nature worship makes use of what is known as tree lore (Harvey, 2007); such lore dictates that trees are responsive to human engagement with them; 'Many Pagans … assert that trees are willing, and able in some way, to communicate things that they would otherwise be unaware of, and therefore the "speaking with" and the "hearing" may also pursue information' (Harvey, 2007: 132).

Communicating with trees may sound primitive, irrational even, to those outside of these forms of ritualistic worship but it is commonplace for many of those individuals involved in alternative spiritual practices. Nature-worship seeks to produce a depth of ecological understanding that is not made possible through the celestial entities of the Judeo-Christian theism that maintains a significantly wider congregation. Magick in this sense is far more tangible (in the mind of the practitioner) than any internalised prayer system might be, requiring not just emotional and spiritual but immersive physical response to a spirituality that exists within the landscape itself. Material offerings left at Ankerwycke demonstrate the site as a place of exchange; ribbons tied in knots are the products of spells; crystals encourage positive energy flows; photographs pay tribute to lost loved ones for whom it is hoped that the tree will keep safe. There is then, a process of trinketisation occurring upon the tree itself, the out of place items initiating a sense of the mobility of the sacred; things are brought to the tree and in doing so are viewed to have become magickally valorised (Foley, 2011; 2013). Through the circulation of material offerings at the Yew, we see a change in the perceived state of the objects which could only be brought about through their movement into the site itself. These trinkets are placed at the tree as payments of love and respect to the spirit of the place, its *genius loci,* and in doing so there will exist the hope that such an act can bring about an affective change to a space outside of the tree (Foley, 2011) – that its preternatural powers will become mobilised. Such a process sees a confirmation of the extended veneration of the sacred space via material souvenirs. Through a bringing to, rather than a taking away of the sacred, 'the more-than-representational qualities which material objects might stand in for' (Maddrell and della Dora, 2013: 1120) augment the sacrality of the tree; the knots, ties and corn dollies which adorn the site do not merely infer a sacredness, they confirm it, tying both practices and people to the tree itself.

Trees and forests have long been considered as sites of strange spiritual occupation, acting as the starting point for myriad fairytales and folklore (Loukaki, 1997; Porteous, 2002; Maitland, 2012). Jones and Cloke (2002) express that the postmodern era has seen a return to the vision of forests as mystical. Woodland has always been envisioned as a place of tension between man, nature and the unruliness of the wilderness: 'forests are imbued with notions of nature, mystery and lack of light, with a clear sense of risk ... of becoming lost, far from home' (Jones and Cloke, 2002: 24). It is no wonder then that those spiritualisms specific to a deeper ecology, turn to the tree as a source of connectivity to the otherworld. Trees and forest are intrinsic to Paganism, referred to as the 'Greenwood', a place away from the domination of humans, a wilderness that 'some steps closer to the Otherworld' (Harvey, 2007: 7).

The links between enchantment, mysticism and the tree-world are by no means a neglected area of study. Work into the mysticism of trees is touched upon in the anthropological work of Rival (2001). Pogue Harrison (1992) places an emphasis upon the symbolism of sacred groves, positing that the forests were the original dwelling place of God, later reflected in the architecture of cathedrals whose arches mirror the cloaking of the canopy. Both Porteous (2002) and Maitland (2012) describe the linkages between folklore, fairytales and the forest insofar as positioning trees within the oral traditions of storytelling. Here the emphasis is on the social place of trees and their being situated alongside people.

Ankerwycke lends itself to the study of mysticism in a more nuanced manner; it is not merely a case of mobile sacrality but additionally offers itself as a way of reimagining the landscape, of reencountering nature as vital (Blain and Wallis, 2004). (Neo)Pagans see the vitality of nature as a source of enchantment (Wallis and Blain, 2003), a view that can be extended beyond mysticism through a framework of mobility; a close rendering of the affective nature of magickal movement. Having provided an introduction to the Ankerwycke Yew, the following section will present a descriptive account of the ritual practices that have been observed at the site and its surrounding area.

Samhain: The Ritual

Samhain (pronounced *sow-ain*) is the name given to the celebration of Halloween or the beginning of the Pagan New Year. It traditionally marks the entry into winter and the end of harvest and is symbolic of the death of the land during this period (Hutton, 2001). The celebration occurs between October 31 and November 1 and is often viewed as a festival of darkness. Samhain 2011 was my first ritual at the Yew and occurred at what is arguably the most important event of the Neo-Pagan calendrical year.

My presence at the Samhain ritual was by no means happenstance. I had come across the Yew a few months earlier whilst out on a Sunday afternoon walk. Within a short time I had stumbled upon the Ankerwycke Yew, which, though unsigned, is located alongside a naturally formed footpath and so is readily accessible to anyone exploring the area. This first sighting of the Yew left me awestruck, initially due to its size and formation but not least because it had been heavily decorated with candles, ribbons, polished stones and flowers. Tree dressing is not uncommon among Neo-Pagan and rural folk customs (Hole, 1976) and this appeared to be an example of something similar. Furthermore, the remains of what was seemingly a ritualised performance lay upon a substantial branch which rested upon the floor to the side of the Yew; melted candle wax, incense and strange flower arrangements had been left in situ, objects of an 'informal memorialisation' that furthered the potential sacrality of the spot (Maddrell, 2013).

I took to the internet to research what the ribbons and charms might mean; a quick search of 'Ankerwycke Yew' and 'pagan' delivered immediate results. The top link was to a directory page entitled UK Pagan Links <www.ukpaganlinks. co.uk> which in turn gave me a description of a Neo-Pagan group named the Circle of Ankerwycke who claimed to hold rituals at the Yew. Better still was the contact details for their 'leader' (who I will refer to here as Sandra), who the page instructed to telephone if one was interested in attending either the moots (monthly meetings of the group) or the rituals. I called the number armed with a list of questions. The discussion with Sandra was brief; she confirmed that ribbons had been left by her group, surprised they were still there as they had formed

Figure 4.2   A candlelit pentagram beneath the Yew
Source: Author's own photograph

part of the February celebration of Imbolc (the close of winter festival). Sandra provided me with a time and date and invited me to attend the Samhain ritual.

The Samhain gathering first took place in a National Trust car park, some quarter of a mile from the Yew. The Circle met at 7.30pm. Night had already closed in and in the absence of street lights, the whole area was now pitch black. Members made their way down through the meadows to a tree lined avenue which would lead to the Yew itself. When approaching the avenue there sounded the unsettling drone of intense rhythmic drumming. Candles flickered through the trees illuminating glimpses of cloaked figures. Only some members were adorned in ritual cloaks, black and purple gowns: others were in casual wear having presumably come straight from work. There was a prevailing sense of anticipation amongst the group; a few individuals were trying to talk above the drumming whilst the majority of the figures in the shadows remained silent. The group paused at the beginning of the arboreal passageway and the drumming stopped. The ritual narrator, half lit by the flickering of candlelight and half hidden beneath his ceremonial hood, spoke of the purpose of Samhain – the importance of celebrating the end of the year, the end of harvest and the end of mortal lives. This was read from his mobile phone, which provided us with a curious juxtaposing of the ultra-modern and (supposedly) ancient rites that were being read. The drumming resumed as we were led down the avenue of trees towards the entrance of a small labyrinth, constructed of ribbon and candles.

Instructions were given to enter the labyrinth, to ponder death as a matter of change or transformation – 'a moving between realms' – and to consider the decisions one had made over the previous year. Entering the maze, it was obvious that there had been orchestrated attempts at providing specific vistas onto the place, so as to illuminate different sides of the sacred tree. This movement towards sacredness was contrived, strategic; systematically devised to heighten the sense of magick at every turn.

At the end of the labyrinth lay the candlelit entrance to the tree itself. The group automatically arranged themselves into a circle. I joined them. Another cloaked man handed a candle to everyone gathered around the tree, followed by the passing of ribbons. The ritual leader spoke of the ancient history of the tree, of our ancestors who struggled through the harvest and of the sacred geometry; the ley line, on which the Yew had grown from and which connects it to a host of other 'magickal' sites across the landscape, a perpetual transference of subterranean energy, a continuous movement of the magickal. Conceived by Alfred Watkins (1921), leylines mark specific path or trackways that join settlements with sites of ritual and sites of trade (Devereux, 1994). These supposed leylines, we were told, connected the group within a network of mobilised spiritual energies.

We were then asked to tie the ribbon we had previously been given, to the right wrist of the person to our left. This was performed in remembrance of someone we knew who had passed to the other side. Some of the members began to weep; the ritual played into something of a grief work – the Yew allowing for a place of mourning. These evocations of bereavement, of both physical and emotional loss, now positioned the site within a 'deathscape' (see Kong, 1999; Maddrell and Sidaway, 2010), a topology of mourning through which the participants could celebrate and connect with the otherworld. The candles were then lit; one by one, acting as illuminated callings for loved ones who had passed; the light of the flame intended to attract spirits on a night, when the veil between this world and the next was at its thinnest. The candles were then placed into a pentangle shaped frame that rested upon the floor (Figure 4.2). This would focus the spirits' attention on the tree place, freeing them from the earth. Traditionally, Yew trees have been planted to commemorate the dead and as Bevan-Jones states in his *The Ancient Yew* (2002), many a graveyard still shelters a yew tree. One of the Circle members had described to me previously how yew trees were planted in graves as living headstones, the mouth of the dead, as Battista *et al.* concur, there are those that believe that spirits can gain access to the world of the living via the roots of trees (Battista *et al.*, 2005).

Praise was then given to the spirit of the place who the circle referred to as *Morwenna Cygnis*, a female deity believed to dwell within the Yew; guarding the tree and protecting the circle. She would later be asked to lead the spirits back into underworld; a symbol of death, the Yew provided a material gateway between this world and the next.

Throughout the ritual, tributes were made to the tree, to its history and to the harvest. Handmade Samhain decorations constructed of twigs, leaves, flora and berries from the local vicinity were placed alongside carved pumpkin lanterns upon the 'altar' – a substantial section of branch that rested by the side of the tree. Cake and mead were offered to the tree in thanks of the fruitful harvest that the circle had been gifted with. Mead was poured onto the base of the tree; it soaked up quickly into the ground as if being drunk by the Yew's roots. The mood was merry as the ritual came to an end, the tree was once again thanked; some individuals placed keepsakes in the nooks of the tree's trunk as offerings, hanging corn dollies and other objects of folk craft from conveniently placed shoots and twigs. This contributed further to the sense of enchantment; strange looking corn figures adorned the face of the tree; making it obvious to outsiders that this was a sacred place. Post ritual; something of a party began. There was much drinking and merrymaking and tales of Samhains past, ghost stories and the like. The tree stood above the circle, watching over the goings on. The gnarled trunk embossed with human-looking faces in the candlelight. The ritual, though now complete, had seemingly brought the tree to life.

Defining Magic(k)

The language of magick can at first appear obtuse. As a rule, magick is the term widely used to describe the practices and rituals encompassed within the occult, things that present as explicitly preternatural (beyond nature) in their origins or affects; spells, curses, unseen agency or in the case of Ankerwycke – spirits. The word magick is generally used within the Neo-Pagan and occult traditions to create a distinction between the magical practices of ritual and the tricks or illusions performed by entertainers; hence magick and not magic. The deviation in spelling is taken from an older form of the word and elevates the performance of the Neo-Pagan over that of the person performing the 'rabbit in a hat' trick (Harvey, 2007). Magick deals not so much with sleight of hand, but with a commanding of the unseen forces of the world.

Given the general sense of intangibility and lack of scientific support that surrounds magick (the two systems being seemingly incompatible), it should come as no surprise that definitions of the term are varied and often rather conflicted; from an embryonic form of science (Crow, 1972) to the transmutation of negatives into positives (Starhawk, 1982) to a transformation or movement of and between things. Aleister Crowley posited that 'Magick is the Science and Art of causing change to occur in conformity with Will' (Crowley, 1976: 131). Crowley's definition inflects the Cartesian model of knowledge and ascribes the notion of magick to a challenging of conformity, a way of igniting disorder or irregularity in an otherwise predictable system. In applying such a concept to nature-worship, Neo-Paganism often seeks to cause a change in the regularity of the ordered world making the ordinarily static become mobile, perform; extending agency to the largely inanimate. Such is the case with the Ankerwycke Yew, which as the account describes, appears to become more and more alive as the ritual proceeds.

Magickal practice enables the individual to enter into a much 'deeper' ecological understanding of their surroundings, a rejection of the anthropocentric in light of an understanding of man existing in and of nature (Naess, 1997), away from a hierarchical system in favour of a world of constituent agents belonging to a multi-faceted universal being. Magick asserts that any one part can move position so as to affect or influence any other part, bringing about a change. What Crowley initiates with his understanding of telekinetic correspondence between things is a mobility of magick.

Another important definition of magick, one that brings us back to nature itself; states that magick is 'at once the knowledge of a network of sympathies or antipathies which bind the things of Nature and the concrete implementation of this knowledge' (Faivre and Needleman, 1992: xvi). Again, the emphasis is on magick being produced via a mobility of knowledge. Contra the abstract conformity that Crowley alludes to, Faivre and Needleman explicitly express magick as the knowledge of Nature's complexity – a complexity that exists to be undermined and manipulated by the practitioner of magick craft. There are telling signs in this account that the magickal is pre-existent with the natural world, that rather than the creator of something new, the practitioner is the impetus for mobilising changes in the 'sympathies or antipathies' which hold together and cohere this natural world. Together these two definitions act so as to elucidate two key facets in our ecological understanding of the surrounding landscape as magickal, and more pivotal still, the description of the Yew during ritual. Firstly, that nature is somehow ordered, regular and open to disturbance (by the magickal practitioner) and secondly, that magick is innate, that it is a mobilisation of the agency that readily exists within all things. The magickal landscape exists within a network of pliable entities, each one waiting for its preternatural agency to be conjured or to be worked upon. Unfolding before us (Wylie, 2002), landscape affords us a space in which to dwell within and upon (Ingold, 2001), an environment to move through, to encounter, to memorialise (Tilley, 1994), it provides the backdrop on which the very practices of magick might occur.

The same equivocation we see in definitions of magick, apply to the practices that belong to it; it is a craft that is practiced with equal multiplicity, however chants, meditation, ritual performance and spell casting are fairly commonplace and are endorsed both privately and publically by the Neo-Pagan community. Each of these practices is used to engage with a place of personal or ritual importance, to invoke gods, goddesses and spirits, to change fortune, to heighten energy or as with the case of the Yew, to commune with nature. In turn all of these practices contribute to the making of a place, through targeted methodologies like the casting of a circle there is a re-imagining of agency, of history and of the affective nature of the landscape. This leads to an initialising of mobility; a movement in objects, energy or spirit. Casting a circle is the practice of creating sacrality; chants, citations and poems are read in addition to the calling of the quarters, the four guardians of the North, East, South and West and the sweeping or cleansing of the area. During the Ankerwycke ritual, we saw the space within the circle formed by the group create a sacred space for the practicing of magick, the performance of ritual and the worshipping of the divine elements (Harvey, 2007: 44). The production of this space was not only a practice of place-making but also of movement. There was both the physical movement of bodies from outside to within the circle, the process of casting the circle, of sweeping it clean, and of inviting entities to move in and between the sacred space that had been formed. All in addition to the believed flows and transference of energy and motions acted out within the ritual itself. What we have then is a place of new extended affective spatiality, one which is liminal, transgressive, existing simultaneously as a threshold between realms of physicality and the incorporeal. The cast circle forms a place for believers, where not only the natural but nature *per se* becomes challenged, complicated and problematised.

Enchanting Place

Where might we position magick theoretically and how does it fit into a critical understanding of the sacred landscape? For this we must look at how magick can be viewed within a conceptual framework of placial mobility, of the peculiar agency of sacred places. Magick is a process of enchantment; it functions as a deeper ecological awareness to bring about willed changes to the natural environment.

Enchantment informs mysticism; it is the process whereby objects of banality become re-envisioned as alive: vital matter. Such a process 'connects across the boundary between the tangible physical world of experience and the intangible, unknowable world of spirit; and these connections, or communications, are fundamental to the everyday worlds of those who experience them' (MacKian, 2011: 69). Magick does not replace nor supersede enchantment theory, rather it mobilises enchantment through a particular set of methodologies. Sites like Stonehenge and the standing stones seen at the village of Avebury, UK become viewed as enchanted because they are inexplicable, despite the workings of archaeologists and historians to unravel their mystery. Such sites as these provide places of ritual, their very mystery – tangible and yet incomprehensible – engendering the perfect gap for mysticism to fill. Magickal ecologies are fostered through sites like these, heterotopic spaces that are not only 'empty vessels' (Coleman and Eade, 2004) where we seek to impart meaning but furthermore spaces like that of the Yew, demand an unorthodox engagement in which to be sacralised, appearing unnatural and then becoming seemingly more so through the alternative spiritualities that they encompass. It is the profound sense of spiritual attachment that is awakened in individuals who visit these sites that appears magickal to them. Through participant observation, Holloway's (2003) analysis of Neo-Pagan traditions describes a process whereby sites are made sacred through practice, their sacrality he posits, is down to the combination of historical and contemporary rituals held there and moreover, the interplay between space and embodiment. Holloway observes that through the relations and interactions between the human and the non-human 'a unity-in-difference is performed through the subject's and other's possession of a unique and qualified agency in relation to one another' (Holloway, 2003: 169). In other words the removal of any hierarchical order that would ordinarily exist between man and nature leads to an extension of agency. But such spaces and rituals rely upon a preconceived notion of the sacred; the monumental nature of the space somehow affords it supernatural qualities. And herein lies the difference between the established sacred and the magickal.

Magick forms sacrality; as we have seen in the casting of the circle, the place is made, sacrality evoked. Magick enables the sacred, it allows it to transcend prescription. Magick does not require a fixed point for its practice; it may use but does not rely on pre-existing sacred spaces, such as Avebury or Stonehenge. Rather, practitioners of magick invoke the spirit of place in order to perform rituals. In other words, magick is a strategy, for both enchantment and for essentialism. Strategic essentialisms are used to construct sacrality; the embodiment of the space by the ritualist not only occurs in particular to this space but it is one which cannot be felt by those outside of the performance. What exists then is a site which is temporarily sacred, one which for a limited time only affords a preternatural agency to N/ nature. I only wish to acknowledge the differences between the magickal and the otherwise sacred, as well as illustrating the emphasis on mobility and ecological understanding that are inherent with magickal practices.

In order to consider enchantment as a process of magick, it is necessary to pay some attention to what enchantment is. The enchantment tale presupposes a meta-narrative of disenchantment, as this much is true; in order to be enchanted, to be filled with wonder, one must be of a world that is essentially banal, disenchanted, but where there is still the potential for the sacred to re-emerge. Enchantment calls forth the spirit of nature, it confronts and reawakens the senses, animating the subject of our gaze; Bennett posits; 'enchantment is something that we encounter, that hits us . . . to be enchanted is to be struck and shaken by the extraordinary that lives amid the familiar and the everyday' (Bennett, 2001: 5). Furthermore, Bennett details such a theorising of enchantment as being constructed of two facets: firstly, an unprocessed feeling of wonder, of being charmed by a novel encounter: secondly, a sense of the *Unheimliche* (uncanny), of being dislocated from one's default sensory perception to encounter something both strangely familiar and other (Bennett, 2001). It is through this definition of enchantment that an application to place will occur. Such a positing of enchantment leaves the purpose and causality of the process open; indeed, Bennett's conception of enchantment is both non-reductionist and non-teleological (Holloway, 2006). The absence of both *telos* and reasoning leaves Bennett's enchantment open to interpretation and reapplication; it is precisely because of the sheer *openness* of this enchantment theory that its attachment to place is unproblematic. As Bennett focuses upon the enchantment of naturally occurring phenomena, from the connectivities of thought to bio-science, a move to the enchantment of place feels both reasoned and logical when describing the production of the sacred mobility at Ankerwycke and the experience of the ritual's participants.

Subscribing to Bennett's enchantment (2001) assumes two things of a place that we might consider enchanted; that it invokes wonder and novelty, and that it creates a feeling of discomfort, of being beyond a normative consideration of natural. Bennett implores the reader 'to keep an eye out for contemporary practices and experiences that are anomalous within a world understood to be wonder-disabled. In other words, to foreground cultural sites that ought not to exist in the way that they do, within a disenchanted world' (Bennett, 2001: 84). One might well consider enchantment synonymous with the notion of sacredness – sacred places are comforting and disturbing, novel and wonderful, 'they are places where the life-force of the Earth wells up' (Harvey, 2007: 146). Sacred places are sacred precisely because of their immediate connection to the fabric of Nature or the divine; they act as gateways to all that lies beyond nature. Furthermore, sacred sites are the points at which man connects to the spirit of nature, heavens, divinities; they stand as the physical markers of man's entanglement with the other or inner worlds and anchor culture to landscape (Ivakhiv, 2001). It is a process of enchantment, of meeting the uncanny and the wondrous simultaneously, that constructs sites of the sacred; the sacred itself then calls for 'anomalous practice', for mysticism, for a movement beyond one's normal thresholds and into the space of liminality through rites or ritual. The enchantment of place then, calls for an act of transgression, a movement to the uncertainty of nature. Places where enchantment is experienced, where the sacred or mystical is transfixed upon a transgression beyond nature, might well be found in sites of magick, ritual or Earth mysteries. Mysticism (and magick) threatens reason and rationality, it holds the potential to confuse reality (Matless, 1991), but this in itself does not negate the truth of experiencing enchantment. Trees, tors, henges, standing stones; all are physical matter and all form part of an embodied mystical landscape; enchantment occurs as a result of such materialities reconnecting us with nature, of allowing the potential for re-imaginings of what nature might be.

Closing Comments

Moving towards a theory of the magickal mobilises the pre-existing components of deep and eco spiritualities, mysticism and *genius loci*. Moreover, magickal ecologies provide a rich and varied landscape on which extremities and anomalies of nature can occur. Nothing transpires outside of nature, there is no outer subjective conscious that can be ascribed to those things that incite the uncanny (Farber, 1967). The magickal provides a space in which to both usher in and analyse phenomena that have previously resided as Fortean, supernatural or paranormal. These previous ways of mobilising 'that which is beyond', do nothing to aid the debates around critical geographies of alterity; rather they further marginalise and segregate discussions of the strange from a discourse of place and spirituality

(MacKian, 2011). Magical ecologies provide an option for getting back into place, for its re-imaginings and for its subsequent animation.

The materiality of the Ankerwycke site, the altar, the decorations, the candles, ribbons, grass, sticks, leaves, branches and so on; these all play a role in the coming to life of the Ankerwycke Yew. To be sure, the sacred cannot be accessed other than through these material engagements with the place, through the offerings left at its corporeal body (the tree) and through placing one's own body within its vicinity. The vitality of the tree, its spirit or agency, is something that is conjured and the ritual performance based at the site works to perpetuate this. When I asked one of the Circle members why they work with the Yew, they replied 'because we are meant to, because the tree chose that place for us'. The same thing might be said of the tree itself; it feels as though it is meant to play host to rituals. That it is a place where our entanglement with nature can be witnessed. The practices that took place during the Samhain celebration helped to mobilise the site, not only in the sense of setting it within a magickal context but also in the way that it began appearing to move, gesticulate and converse with the Circle.

The movement in and around the sacred circle beneath the tree illustrates the importance of mobility; in defining magick we saw two common elements, transference of energy and transference of knowledge. The rituals acted out under the tree honour this; they mirror movement in myriad forms; through procession, dance, light, chanting, energy. When thinking of the Yew, what occurs to me is that it is not merely a historical lineage that engenders sacrality here, nor is it just the size and scale of the tree that provokes enchantment. Rather, it is the vitality that the ritual plays privy to which informs the enchantment process. Nothing particularly 'magickal' occurs. Yet, there is a privileging of the agency of the tree that does not exist in the vernacular for the majority of us. The enchantment of the Yew comes about as a direct result of the interplay between its fixity and the ongoing processes of mobility that occur around it. The treatment of the tree as 'magickally' sacred rests upon this practice, both in terms of its mobility and of its becoming enchanted, an extension of the affective qualities of the Yew that sees it shift between the everyday and the more-than-representational. It is afforded essence, raised beyond the strata of the merely non-human. We do not readily come across sites that are so obviously ancient as this and yet rarer still, so bizarrely formed. Furthermore, the treatment of the Yew as alive; the manner in which it composes itself in the candlelight; the ways in which the landscape seems to have adapted to produce such a space as this, all heighten the feeling that this is how the place is supposed to be; that nature has somehow intervened in the order of things. If these feelings exist only for the purpose of the ritual then this does not in itself deny essentialism, rather it points to the idea of the essential being used strategically; that it exists temporarily, to be embodied by the place for the purposes of sacrality and the practicing of magick. Magick is about understanding the world at both micro and macro levels; its user seeks to affect and be affected by the occluded energies of the earth. Magick is a mode of mobilisation and is the resulting composite of a spiritually inflected ecological philosophy.

To say that the tree is fixed, rooted or anchored to the site is but to speak of only half of its being. To be sure, the tree is far from static; it grows, develops, reaches upwards and outwards, gathering place for itself. The essence of the site, its supposed unique character or spirit thus develops simultaneously. As the tree usurps earth and air from the surrounding flora, it moves, at a slow but steady pace. The rituals practiced below the Yew's branches play mimicry to such movement; the slow unwinding journey through the labyrinth, the diffusion of the mead through the soil, the subterranean energy line, the transference of knowledge in the rites, of narratives, poetry and song; the procession, dancing, drumming, the movement from darkness into light, from death to life and back again. The magick of the tree comes about via a process of enchantment that is reliant not so much upon the conjectured fixing of the tree and therefore its energy to one place, rather it is about cycles, rhythms and journeys into the sacred, the materiality of the tree acting as a portal; a gateway to enchantment that stands at the threshold of the sacred.

References

Battista, K., LaBelle, B., Penner, B., Pile, S. and Rendell, J., 2005. Exploring 'an area of outstanding unnatural beauty': a treasure hunt around King's Cross, London. *Cultural Geographies*, 429(12): 429–62.

Bennett, J., 2001. *The Enchantment of Modern Life: Attachments, Crossings and Ethics*. Princeton, USA: Princeton University Press.

Bevan-Jones, R., 2002. *The Ancient Yew.* Cheshire, UK: Windgather Press.

Carr-Gomm, P. and Heygate, R., 2010. *The Book of English Magic*. London: John Murray.

Blain, J. and Wallis, R., 2004. Sacred sites, contested rites/rights: contemporary Pagan engagements with the past. *Journal of Material Culture*, 9(3): 237–61.

Coleman, S and Eade, J., eds, 2004. *Reframing Pilgrimage: Cultures in Motion*. London, UK: Routledge.

Cresswell, T., 2002. Introduction: Theorizing place. *Thamyris/Intersecting*, 9: 11–32.

Cresswell, T. and Merriman, P., eds, 2011. *Geographies of Mobilities: Practices, Spaces, Subjects*. Farnham, UK: Ashgate.

Crow, W.B., 1972. *A History of Magic, Witchcraft and Occultism*. London, UK: Abacus Press.

Crowley, A., 1976. *Magick in Theory and Practice*. London, UK: Dover.

Devereux, P., 1994. *The New Ley Hunter's Guide*. Glastonbury, UK: Gothic Image Publications.

Eade, J. and Sallnow, M., eds, 1991. *Contesting the Sacred: The Anthropology of Christian Pilgrimage*. London, UK and New York, USA: Routledge.

Escobar, A., 2001. Culture sits in places: Reflections on globalism and subaltern strategies of localization. *Political Geography*, 20(2): 139–74.

Faivre, A. and Needleman, J., eds, 1992. *Modern Esoteric Spirituality*. London: SCM.

Farber, M., 1967. *Phenomenology and Existence: Toward a Philosophy within Nature*. London, UK: Harpertorch Books.

Foley, R., 2011. Performing health in place: The holy well as a therapeutic assemblage. *Health and Place,* 17: 470–79.

Foley. R., 2013. Small health pilgrimages: Place and practice at the Holy Well. *Culture and Religion*, 14(1): 44–62.

Harvey, G., 2007. *Listening People Speaking Earth: Contemporary Paganism*. 2nd ed. London, UK: Hurst and Company.

Hole, C., 1976. *British Folk Customs*. London: Book Club Associates.

Holloway, J., 2003. Spiritual embodiment and sacred rural landscapes. In: P. Cloke, ed., 2003. *Country Visions*. Harlow, Essex, UK: Pearson Education Limited, 158–75.

Holloway, J., 2006. Enchanted spaces: The séance, affect, and geographies of religion. *Annals of the Association of American Geographers*, 96(1): 182–7.

Hutton, R., 2001. *The Stations of the Sun: A History of the Ritual Year in Britain*. Oxford, UK: Oxford University Press.

Ingold, T., 2001. *The Perception of the Environment: Essays in Livelihood, Dwelling and Skill*. London, UK: Routledge.

Ivakhiv, A., 2001. *Claiming Sacred Ground: Pilgrims and Politics at Glastonbury and Sedona*. Indiana, USA: Indiana University Press.

Jones, O. and Cloke, P., 2002. *Tree Cultures*. Oxford, UK: Berg.

Knight, C., 1845. *Old England: A Pictorial Museum*, London, UK: James Sangster and Co.

Kong, L., 1999. Cemeteries and Columbaria, memorials and mausoleums: Narrative and interpretation in the study of deathscapes in geography. *Australian Geographical Studies*, 37: 1–10.

Loukaki, A., 1997. Genius Loci?: Contrasting interpretations of the 'Sacred Rock of the Athenian Acropolis'. *Annals of the Association of American Geographers*, 87(2): 306–29.

MacKian, S., 2011. Crossing spiritual boundaries: encountering, articulating and representing otherworlds. *Methodological Innovations Online*, 6(3): 61–74.

Maddrell, A., 2013. Living with the deceased: absence, presence and absence-presence. *Cultural Geographies*, 20(4): 501–22.

Maddrell, A. and della Dora, V., 2013. Crossing surfaces in search of the Holy: Landscape and liminality in contemporary Christian pilgrimage. *Environment and Planning A*, 45: 1105–126.

Maddrell, A. and Sidaway, J., eds, 2010. *Deathscapes: Spaces for Death, Dying, Mourning and Remembrance*. Farnham, UK: Ashgate.

Maitland, S., 2012. *Gossip from the Forest: The Tangled Roots of Our Forests and Fairytales*. London: Granta.

Massey, D., 1994. *Space, Place and Gender*, Cambridge, UK: Polity Press, 146–56.

Matless, D., 1991. Nature the modern and the mystic: Tales from early twentieth century geography. *Transactions of the Institute of British Geographers*, 16(3): 272–86.

Naess, A., 1997. Deep ecology. In: C. Merchant, ed., 1997. *Ecology*. New Jersey: Humanities Press. 120–24.

National Trust, 2010. *The Ankerwycke Yew at Runnymede*. [online] Available at: http://www.nationaltrust.org.uk/main/w-chl/w-countryside_environment/ w-woodland/w-woodland-heritage_trees/w-woodland-heritage_trees-ankerwycke_yew.htm [Last accessed: 04 June 2010].

Pogue-Harrison, R., 1992. *Forests: The Shadow of Civilization*. Chicago, USA: Chicago University Press.

Porteous, A., 2002. *The Forest in Folklore and Mythology*. New York: Dover Press.

Rival, L., ed., 2001. *The Social Life of Trees: Anthropological Perspectives on Tree Symbolism*. Oxford: Berg.

Starhawk., 1982. *Dreaming the Dark*. Boston: Beacon Press.

Tilley, C., 1994. *A Phenomenology of Landscape: Places, Paths and Monuments*. Oxford, UK: Berg Publishers.

UK Pagan Links (2007). *The Circle of Ankerwycke*. [online] Available at: http:// www.ukpaganlinks.co.uk/article.php?n=112andpage=1 [Last Accessed 18 June 2013].

Wallis, R.J. and Blain, J., 2003. Sites, sacredness, and stories: Interactions of archaeology and contemporary paganism. *Folklore*, 114(3): 307–21.

Watkins, A., 1921. *Early British Trackways, Moats, Mounds, Camps and Sites*. UK: Resurrection Press.

Wylie, J., 2002. An essay on ascending Glastonbury Tor. *Geoforum*, 33: 441–54.

Wylie, J., 2007. *Landscape*. London, UK: Routledge.

Chapter 5

'At Least Once in a Lifetime': Sports Pilgrimage and Constructions of the TT Races as 'Sacred' Journey

Avril Maddrell, Alan Terry, Tim Gale and Simon Arlidge

Introduction

The iconic image of medieval pilgrimage is one of arduous journey, often made deliberately onerous in order to enact penance for sins (Digance, 2006). However, pilgrimages also incorporate important aspects of celebration, community and renewal (Dyas, 2004; Maddrell, 2013). Within interdisciplinary pilgrimage studies, Coleman and Eade (2004) have brought the mobilities perspective to the fore. In so doing, they repositioned pilgrimage primarily as ritual and performance, whereby practice serves to sacralise places, rather than the sacred being inherent to a given shrine or site. The analytical purchase of this argument varies according to the specific characteristics of any given example of pilgrimage studied, not least in terms of the beliefs held by practitioners. What is certain is the value of a) examining pilgrimage practices through the lens of the mobilities discourse and b) the need for analysis of the nuanced interplay of journey, place and performance for any individual or collective experience of what is deemed to be pilgrimage. Understanding the range and character of the moto-mobilities particular to motorcyclists also merits further scholarly attention (Pinch and Reimer 2012; Terry *et al.*, forthcoming).

While the Western Christian presentation of pilgrimage as simultaneous physical and inner *journeys* is common to the practices of other faiths, such as Tibetan Buddhism, it does not represent all Christian traditions, yet alone all religious pilgrimage. Nonetheless, whatever the form, any discussion of pilgrimage tends to presume a combination of religiously-motivated travel and ritual. However, in the twenty-first century the concept of pilgrimage is no longer limited to those journeys motivated and sanctioned *only* by institutionalised religions. Rather, in keeping with more open understandings of what is sacred (Milton, 2002), scholars have expanded the meaning of pilgrimage to incorporate 'undertaking a [any] journey that is redolent with *meaning*' (Digance, 2006: 36, our emphasis). As illustrated in the previous chapters, and highlighted in the Introduction, pilgrimage is a complex phenomenon, with numerous definitions and varying associated practices (Gemzöe, 2012; Reader and Walter, 1993). Most representations and designations of 'pilgrimage' recognise that some sense of the pursuit of what is deemed 'sacred' is crucial. As with religious pilgrimages, ultimately pilgrims self-define (Morinis, 1992), and the language of pilgrimage is increasingly being applied to other secular purposive journeys, which reflect world views, passions and devotion. Such pilgrimages can have enormous range, including war graves (Walter, 1993) and celebrity graves, such as Elvis' burial place at Graceland or Marilyn Monroe's grave in Los Angeles; as well as journeys 'home' made by migrants or exiles (see Shakkour, this volume); and 'once in a lifetime' visits to natural wonders of the world such as the Great Barrier Reef, Grand Canyon or Antarctica.

For some sports enthusiasts this broad sense of sacred undertaking includes journeying to attend nationally or globally significant sporting venues or events (Stausberg, 2011). The Tourist Trophy (TT) motorcycle races in the Isle of Man are one such sporting event, which is frequently represented as a 'pilgrimage'. In this chapter we analyse spectator accounts of their attendance at the TT races, to ascertain the extent to which the TT is discursively constructed and performatively experienced as a sacred journey, place-as-shrine, community and/or form of ritual practice. Other commentators have illustrated how sport can be configured as pilgrimage. For example, in his review of religious tourism, Stausberg notes how the term has been applied to sporting and other endeavours:

> There are 'pilgrimages' to sports and arts events and places. Accordingly, the word Mecca, a prototypical pilgrimage site, is colloquially used to designate a range of special sites such as 'the Mecca of fishing', 'the Mecca of diving', 'the Mecca of speed', 'the Mecca of gaming' or 'the Mecca of Jazz'. The examples are innumerable. And the rhetoric is ambivalent, since some may use the word metaphorically, as if to emphasise the importance of their trip, while others will use it performatively in order to frame the trip as a special event. (Stausberg, 2011: 54)

One example which Stausberg highlights is that of St Andrews in Scotland. With its reputed relic of St Andrew, the town was the second most important pilgrimage site in medieval Christian Europe after Santiago de Compostela (Tobert, 2001; cited by Digance, 2006), but in the present day it is more likely to be deemed a site of 'pilgrimage' for golfers to visit – and especially to play on – the Royal and Ancient golf course. Zakus (2002; cited by Stausberg, 2011) claims that for golfers the significance of this 'once in a life time' experience of playing on the original golf course must be seen in the same way as a pilgrim visiting Mecca. Needless to say, devout Muslims would dispute that this playing of a game has the same significance as the hajj, which is one of the five pillars of Islam, and a religious obligation for all those able to undertake the journey. What *can* be asserted is that some people express their passion and commitment to their sport rhetorically and metaphorically through their references to Mecca as the archetypal example of pilgrimage to a sacred place. They also describe a sense of compulsion to visit the places which they identify as the historical or contemporary 'spiritual' home or pinnacle of their sport.

This representation of visiting sporting venues as a form of pilgrimage has occurred since the early twentieth century, as seen in the case of Caspar Whitney's (1895 [1923]) *A Sporting Pilgrimage: Riding to Hounds, Golf, Rowing, Football, Club and University Athletics*. More recent publications include *Emerald Fairways and Foam Flecked Seas: A Golfer's Pilgrimage to the Courses of Ireland* (Finnegan, 2007), with similar volumes covering Scotland, and England and Wales. These volumes represent pilgrimage as a systematic journey of 'collection', requiring visiting and playing golf on each course in Ireland, but is also, more pragmatically, a rhetorical device to justify and give greater status to the undertaking than any book simply titled 'a guide to' or 'compendium of' the golf courses of Ireland *et cetera* could hope to do. This form of 'pilgrimage' can be seen to have parallels in other sporting 'collections' such as 'Munro Bagging' for hill walkers in the Scottish Highlands. In some cases this process of 'collection' affords the opportunity to emulate sporting heroes, to 'walk in their footsteps' and measure one's own skill against professional performances. This may indicate a desire, on the part of at least some sporting pilgrims, to experience something of the challenges and embodied experiences of their heroes in situ (a theme that will be returned to in relation to the TT races).

Thus, in secular contexts, purposive journeys motivated by the desire to visit specific sporting destinations, either sporting venues or places associated with sporting heroes, have been represented rhetorically as 'pilgrimages' for over a hundred years. Sporting pilgrimages that centre on places associated with sporting icons typically include their place of birth, grave and sites of key sporting achievements. The 'Australian Sporting Pilgrimages' website is dedicated to such itineraries, including one in honour of cricketer Don Bradman, described as 'the greatest Australian batsman of all time' (Whitley, n.d.). The Bradman 'pilgrimage' has a number of key sites with which he was associated, including Bowral, New South Wales, where he grew up; the Bradman Oval, named for him; and the place where his ashes were scattered. Other sites include the Adelaide Oval where he made his first class debut and scored his highest innings and the State Library of South Australia which houses numerous memorabilia associated with Bradman (Whitley, n.d.). This focus, not only on places but also on memorabilia and artefacts, has parallels with the veneration of relics at many sites of religious pilgrimage.

Interestingly, while the focus for sporting veneration is commonly a track, course or other site of ideal practice – that is, a place designated as sacred – in some cases the sacred qualities may be attributed to an individual person, as illustrated in Wojcik's (2008) study of the late USA track athlete Steve Prefontaine. Often referred to as 'Pre', Prefontaine was admired during his lifetime for his inspirational commitment to his sport and his embodiment of the 'American dream'. After his death in a road traffic accident in 1974, it was the site of his *death* which became sacralised, through the repeated acts of remembrance by fans and the material offerings they left at the site (Wojcik, 2008), illustrating Coleman and Eade's (2004) thesis of sacralisation by performance. The continued practices of successive generations of runners has ritualised individual and group visits to the site, exemplified by youth team visits led by coaches and the many who leave personal and valued objects, especially those associated with running, at the memorial, such as competition numbers, shoes worn during a race win and even medals. Pre's values and status are reinforced by DVDs on his life and career, and the status of the site of his death at Eugene, Oregon, has been formalised with the provision of parking spaces and a permanent memorial. While one track coach described Eugene as a 'Mecca for running', and a local track has been renamed in honour of Pre, his roadside memorial also attracts ritual behaviour and communication with the deceased, including supplications (Wojcik, 2008). These are acts which typically characterise religious pilgrimage, and memorials where the dead are credited with absence-presence (Maddrell, 2009, 2013). Yet here 'People travel to the site not only for reason of commemoration or curiosity but to receive personal benefits in the form of inspiration, aid, or blessings, and they speak of feeling renewed or transformed as a result of their visit' (Wojcik, 2008: 218): that is, visiting Pre's Rock is an acknowledgement of the spiritual side of running and, for some, a spiritual experience in its own right. Thus the deceased Pre - interestingly at the site of his death rather than the site of his burial - is credited with the talismanic qualities more typically ascribed to saints or other powerful mystic relics. Another example of homage to late sporting heroes being portrayed as pilgrimage was reported by *The Sun* newspaper, whose headline read 'Jose Maria Olazabal is considering a pilgrimage with the Ryder Cup to the final resting place of his best friend Seve Ballesteros' (Dillon, 2012). This expresses a sense of connection between living sports stars and their deceased colleagues, and a desire to share and possibly attribute their success to the deceased.

What is clear from these various examples is that 'in everyday language and publicity, pilgrimage tropes can be employed rhetorically in order to mark an experience, a place or an attachment as potentially extraordinary' (Stausberg, 2011: 54); this includes iconic sites associated with particular sports or sporting heroes. As indicated by the wide range of forms of secular pilgrimage discussed above, many tourists find 'transcendent meaning' in secular places and practices they deem to be significant; for example, those visiting Niagara Falls describe themselves as 'pilgrims' and use 'Religious tropes ... to describe places and experiences' (Stausberg, 2011: 14), as do those visiting war graves, especially those associated with family members (Reader, 1993). However, this discursive framing of travelling to important iconic sporting destinations is particularly common; and the combination of pilgrimage and mobilities literature as frames of reference allows us to interrogate the meaning and significance of these journeys and experiences (see Cresswell and Merriman, 2011). We will now address how this relates to motorcycling and, more specifically, spectator experience of the TT races in the Isle of Man.

Biking as Pilgrimage

Motorcyling is relatively under-studied as a form of mobility and tends to be associated with youthful and rebellious culture (Pinch and Reimer, 2012). Hence, for most people, images of motorcycles are unlikely to spring to mind when thinking of pilgrimages, but different forms of cycling, including motorcycling, are occasionally used as part of religious pilgrimage. Cycling, along with horse-riding and walking, is a mode of travel recognised by the Roman Catholic pilgrimage authorities of the Camino de Santiago de Compostela in Spain. Cycling also occurs in the annual feast day of Ħaż-Żabba, in Malta, which has featured a motorcycle and bicycle pilgrimage since 1957, Our Lady of Grace, the town's patron saint, being known as the patron of cyclists. But perhaps the best known religious *motorcycle* pilgrimage is that of La Madone des Motards (the Madonna of the bikers) held every August since 1979 in Porcaro, Brittany. Each year up to 10,000 motorcyclists travel from across Europe for a service of blessing by a Roman Catholic priest followed by a 45-mile pilgrimage ride, which has developed into a wider motorcycling festival. In addition to these religious rituals, some holiday companies also describe their secular motorcycling tours as 'pilgrimages', either due to an itinerary which follows a pilgrimage route, such as the Camino de Santiago de Compostela, or through the use of the term 'pilgrimage' as a discursive frame to reflect the ways in which the excursion is represented as a journey of self-discovery or annual ritual.

Jill Dubisch, a leading scholar on religious pilgrimage, describes how she and her partner came to understand their experience of the 'Run for the Wall', a collective 'ride-out' across the USA to the Vietnam memorial in Washington DC, as a 'pilgrimage'. For Dubsich (2004, 2005, 2008), a number of characteristics and social dynamics justified the term: the long journey to a sacred end-point; the sense of liminality; the sense of solidarity or 'brotherhood' [in what was largely a male group]; the sense of compulsion to participate; and the therapeutic visceral experience of motorcycling. Furthermore, these elements 'combine to create a psychological receptivity to the ritual messages imparted along the journey's route' (Dubisch, 2004: 106). These combined factors clearly demonstrate significant common ground with the community and purpose of faith-based pilgrimage. The 'Run for the Wall' has a sacred destination (the Wall as memorial, sacred to the dead) and purpose (continuing the political battle to 'bring home' those still missing in action or buried in Vietnam), 'but also combines the individual search for healing and identity with the creation of a collective narrative' (*ibid.*), all of which shape the individual and collective journey. Thus, 'The Run' has become an annual event with its own practices and rituals. For many participants, personal healing is sought and the term 'pilgrimage' is often used by participants during the journey, with none of the participants seeing themselves simply as 'tourists': 'We're not tourists. We're pilgrims' (Dubisch, 2004: 113; 2005, 2008).

Moreover, Dubsich argues that motorcycling itself can be a spiritual experience (as can be the case with other sports, illustrated by Wojcik's (2008) account of runners' rituals and experiences above). In the case of motorcycling, each rider is enclosed in their:

> ... own space of noise and movement while simultaneously being part of the larger formation that visually (and audibly) represents the collective endeavour that is 'The Run'. In this sense, the road itself and the liminal experience of riding – becomes a sort of shrine, a space and place beyond the ordinary in which altered states of being may be experienced ... that contributes to their receptiveness to the experiences, impressions, and the messages encountered in the course of the journey. (Dubsich, 2008: 307–8)

These observations were echoed by Reverend John Caldwell, the vicar whose church hosts an annual service for those at the TT races and who is a keen motorcyclist himself.

> Motorcyclists are quite spiritual people in their own way ... One of the things I always liken it to a dog that sticks its head out of a car window. It is the same experience riding a motorbike. It is this free spirit thing. ... . when you are on a bike with a crash helmet on, visor down, you are in your own space. And you are in touch with you [yourself] ... the human race has lost its ability to be in touch with itself because of the busyness of modern life ... inside a crash helmet you can still withdraw into your own space. That is one of the dimensions to [motorcycling] that people don't pick up on, [that bikers] are spiritual people. How they pick up on that spirituality is through their bikes, whether [or not] they are Christians. (Interview, Reverend John Caldwell, Isle of Man, 2010)

These comments on the embodied experience of biking, the sense of personal space and reflection whilst riding, and commitment to mutual support, represent an ideal of motorcycling experience and community which underscore and affirm Dubisch's observations made in a more formal context.

What of the TT races? Can travelling to the TT be considered a 'sacred journey'? We now turn to the specific experience of the TT races in the Isle of Man in order to consider the extent to which practices, rituals, metaphors and experiential narratives are deployed to frame this place-based event as 'sacred' and a form of pilgrimage.

The Isle of Man Tourist Trophy (TT) Motorcycle Races

The Isle of Man TT motorcycle races date from 1907, and quickly became dubbed as 'the most important sporting event on the calendar of the international motorcycling community' (Disko, 2007: 3). The circuit is made up of just under 38 miles of closed public roads, with races ranging between 4 to 6 laps, and the fastest riders lapping the circuit at average speeds over 130 mph. Hence 'the combination of its length and technical challenges have resulted in the TT being known as the ultimate challenge in motorcycle racing' (Terry *et al.*, forthcoming), with significant risk of injury or death for competitors (Fern, 2003; Watterson, 2010). Despite being axed from the Motorcycle Grand Prix World Championship in 1976, each year the combination of risk and skill (coupled with the wider festival and the island itself) attracts between 31,000 and 38,000 visitors to the TT races accompanied by some 10,000 motorcycles (Isle of Man Treasury, 2011).

The findings we discuss in this chapter are drawn from a wider study of the TT races undertaken by the authors during TT Practice Week in 2010 and Race Week in 2011. One author grew up on the Island, while the others were first time visitors, two of whom are daily motorcyclists and travelled to the Island on their bikes. The study used mixed methodologies including interviews, questionnaires, word association sheets and participant observation. All of these are drawn on in the discussion below, but the main data source referred to here is the questionnaire, completed by 167 randomly selected spectators at the races. Each questionnaire respondent cited below is given an identification number and details of visitor/resident status, sex and age group (for example, 146, Resident, Female, 30–44 years). Eighty-one per cent of the sample comprised visitors and 19 per cent were residents of the Isle of Man. Two-thirds were over 30 years of age and 79 per cent identified as male; and none of the sample included visible ethnic minorities, all of which reflected the observed demographics of the event (Field Diary Notes): essentially TT spectators are predominantly white middle age men. While the selection of responses cited below are largely determined by theme, we have endeavoured to include representation of the minority cohorts (for example, women and international visitors who make up 21 per cent and 10 per cent of respondents respectively), and will reflect on any implications relating to different sub-cohorts in the following analysis. The responses discussed below are largely drawn from an open question asking spectators about what motivated them to attend the TT festival; no questions were asked about the 'sacred' qualities of the TT races or their experience.

The TT as Sacred Journey

As noted above, biking can be a 'spiritual experience', in part relating to the embodied experience melding riding and reflection within the 'bubble' of the helmet, and in part because of a constant background awareness of risk and mortality when riding (interview Rev. John Caldwell, 2010). A few bikers do attend the TT 'Mad Sunday' and other church services during the festival, and competitors and their families are known to request prayers from the TT chaplain (*ibid.*). However, in this chapter we want to focus on examining a wider notion of the TT as 'sacred journey', rather than spiritual practices at the TT races per se.

Representation and Metaphors of TT as Pilgrimage

Spectators are often represented as passive observers at sporting events, but in fact have their own significant mobilities (Terry *et al.*, forthcoming), not least in the case of the TT where non-residents have to cross the Irish Sea in order to reach the Island, and many travel hundreds of miles before reaching the port or airport, including those who travel internationally by motorcycle (our sample included motorcyclists from Belgium, Italy, Germany and France as well as air-travellers from the USA, Canada, Australia, New Zealand and Japan). The majority of spectators also bring or hire vehicles to allow movement to and from favoured stretches of the long race course and to explore the rest of the Island on non-race days, with two thirds favouring the scenic south west of the Island for touring and recreation.

The journeys made to reach the Island, especially the more arduous ones, contribute to visitors' sense of undertaking an embodied journey to a 'sacred' destination, as does the promotional rhetoric that: 'There is nothing on Earth quite like the Isle of Man TT races' (Isle of Man TT, n.d.). This is reinforced by spectators' representation of the TT races and the Island itself as 'sacred'. As with other forms of sports 'pilgrimage' discussed above, the name 'Mecca' was deployed as a simile by TT spectators to signify what they saw as the ultimate status of the races and their desire to attend. One respondent explained: 'Like Muslims go to Mecca, motorcycle enthusiasts go to the Isle of Man' (132, Visitor, Male, 60–74 years), a view echoed by others including international visitors. One woman noted the *long-standing* nature of that view: 'historically the island is known as the Mecca of Motorcycling – one of those 'must do' events' (142, Visitor, Female, 45–59 years). This sense of event heritage is also stressed on the official TT website: 'No other motorsport event can boast more than 100 years of such illustrious history, rich in tradition and legends' (Isle of Man TT, n.d.). These legendary qualities were also likened to the challenges and risks associated with the races: 'the history and the bravery' (35, Visitor, Male, 30–44 years), a view once more reinforced by the TT website which asserts that 'to have your name inscribed on a TT trophy is to sit with the gods' (Isle of Man TT, n.d.), thereby attributing sacred status, within the sport, to the TT champions.

A number of spectator respondents attending the TT festival described themselves as being *compelled* to go to the TT races at least once in their lifetime, that is, self-defining themselves as 'pilgrims', echoing the sense of religious obligation to undertake pilgrimage held by some faiths or denominations. One retired international visitor, attending for the first time, explained that he had 'Wanted to see it since my youth' (37, International Visitor, Male, 60–74 years), his visit being the fulfilment of a life-long ambition. Others explicitly used the term 'pilgrimage' to describe their experience of the TT races. For example, a female visitor described visiting the Island specifically as a 'pilgrimage' for TT supporters; and a 30–44 year old male respondent characterised it as 'like pilgrimage' (117, Visitor, Male, 30–44 years), because he had attended the races for eight consecutive years, so was part of the annual rhythm or patterning of his life.

Others represented the perceived sacred qualities of the TT races in more secular terms, especially through representing the competition itself as the metaphorical pinnacle of motorsports: 'Greatest test of man [*sic.*] and machine' (70, Visitor, Male, 30–44 years); 'It's the Everest of motorsport and spectators share in the experience with the racers' (16, Visitor, Male, 30–44 years).

From 'Once in a Lifetime' to 'Every Year': TT Spectator Longevity and Recurrence

The TT is well known for its repeat visitors (68 per cent of respondents reported previous attendance, a finding confirmed by a 2010 TT visitor exit poll (Isle of Man Treasury, 2011)). Furthermore, within that majority of returnees there is a cohort of supporters whose loyalty to the event prompts *annual* attendance, making the TT not only part of their annual cycle, but a major commitment of holiday/leisure time and finances. For these respondents motorcycle racing and the TT was closely tied to their identity and self-representation, as well as family practices and traditions. *The Manx Independent* newspaper (2011) reported one TT spectator who had attended for a remarkable 64 of his 66 years, but similar patterns of annual attendance were not unusual. For example, one woman reported she had attended the TT for 46 years, 'Every year since I was born' (123, Visitor, Female, 45–59 years). She explained that her father had raced side cars in the TT for 26 years, but clearly what had started out as an event-focused family holiday had been continued by choice and adopted as part of her own adult identity, as she continued to travel to the TT races every year. Another respondent in the 30–44 years of age category reported attending a total 30 times (7, Visitor, Male, 30–44 years), indicative of a family tradition, which he was now continuing with his wife. A retired spectator described the TT races as 'addictive' with almost 30 visits between 1952–2011 (131, Visitor, Male, 75+ years). Others reported attending 15 (11, Visitor, Male, 60–74 years) and 25 times (22, Visitor, Female, 45–59 years), and a fruit farmer from Kent had attended 'Most years since 1973', the TT races having been his annual holiday (130, Visitor, Male, 75+ years). Structural factors such as the limited availability of accommodation on the Island play their part in encouraging recurring bookings, as it prompts visitors to make reservations year on year, often before leaving the Island at the end of the current TT festival (Interview, Mike Doherty, Isle of Man Department of Economic Development, 2010). However, this pragmatic reflex needs to be matched by a passionate commitment to the races in order to attend on a recurring basis when transport and accommodation prices are at their peak. Furthermore, the terminology used here to describe these recurrent practices is suggestive of deep-felt commitment and a degree of ritual and it is to ritual practices that we now turn.

The TT Races: Communitas and Co-presence

In addition to references to fulfilment of dreams and life-long ambition to attend the races, many questionnaire respondents described the TT festival as 'unique' and even 'magical'. Part of this sense of enchantment emanates from co-presence (discussed below) and part of it from the liminal qualities of the 'time-space' or place-temporality (Wunderlich, 2010) of the TT festival, situated as it is on an Island in the middle of the Irish Sea. Attributes of 'Islandness', such as topography, distance of separation, social relations and forms of mobility (Vannini and Taggart, 2012), all contribute to give visitors the sense of a very particular place-temporality for the TT races as separate from the rest of their everyday lives. In addition to these geographical attributes, the political geography is also significant: the Island's independent government, with its willingness to close public roads for the races and its facilitation of a motorcycle-centric space for the duration of the TT festival, is an important agent in creating the particular time-space of the TT phenomena. 'The Island becomes a motorcycle-focused social and competitive space, shaped by its 'islandness' and its independent legislature' (Terry *et al.*, forthcoming). These attributes of 'islandness' clearly resonate with anthropological discourses of the liminal qualities associated with pilgrimage, characterised by ritual, renewal, set-apart time-spaces, suspended status and *communitas* (Turner and Turner, 1978). Whilst the Turnerian model of ideal pilgrimage has rightly been critiqued by subsequent scholarship, the concept still has some analytical purchase (Eade, 2000); these characteristics are nonetheless useful as a framework for exploring the extent to which visiting the TT races is represented and experienced as a sacred journey and experience.

Friendship, a strong sense of community, and a commitment to mutual aid are attributes many motorcyclists share with faith groups and pilgrims (see Hopkins *et al.*, this volume).

> One [other] thing you find within the motorcycle fraternity that you don't always get in broader life is the ability to help each other. If you were riding a bike and you had broken down at the side of the road, I guarantee the first person to stop is a motorcyclist to see if you were ok. That aspect is very strong throughout the motorcycle fraternity. (Interview, Reverend John Caldwell, Isle of Man 2010)

In response to a TT word association sheet used as part of the data collection for this research, 32 per cent of visitor respondents highlighted this sense of 'biker community' as a key attribute of their TT experience, suggesting that both co-presence and shared interests and ideals helped shape that experience. Each race and practice requires over 400 marshals around the course as well as other volunteers such as first aiders, and this 'army' of race-related volunteers is one example of TT-based community, voluntarism and mutual aid. As one respondent commented, 'It feels like one big family affair' (17, Visitor, Male, 45–59 years). The notion of co-presence captures the relational qualities of experience that are only possible through being in situ with others at a given moment (see Figure 5.1). At the TT races the qualities of co-presence are shaped by residents, other spectators and competitors as well as the embodied sensory experiences of the kinetic, haptic, aural and oral phenomena of the noise, speed, excitement, food,

Figure 5.1   The pleasures of co-presence and being in situ
Source: Avril Maddrell's own photograph

drink, crowds, petrol fumes and sea air that constitute the TT festival. Co-presence and the combined liminoid qualities of the Island and TT event as a space and time set apart from everyday social norms and practices and governed by different laws, can result in 'group think' such as the normalisation of speed and risk (Terry *et al.*, forthcoming).

Whilst a sense of community, through family, friends and shared interests, is central to the 'value-added' qualities of actually being in situ at the TT races, rather than watching them on television, another frequently cited highlight was the opportunity to be close to and even interact with race competitors. In the first place, spectators' proximity to the track provides a sense of connection with competitors as individuals: 'Riders are accessible – you can see the faces under the helmets' (22, Visitor, Female, 45–59 years); and secondly, public access to the competitors' paddocks allows informal interaction with competitors: 'Public and racers and bikers all mixing together' (44, Visitor, Male, 45–59 years); 'More accessibility to general fan base for example in pits and paddock during event and more fringe events' (11, Visitor, Male, 60–74 years). This affords spectators the opportunity to actually *meet* their heroes, an experience frequently cited as the high point of their TT visit: 'Better [than other races] can speak to riders/crew during event' (115, International Visitor, Male, 60–74 years); 'Meeting Giacomo Agostini in the pits and getting his autograph' (142, Visitor, Female, 45–59 years); 'Seeing DJ [Dave Jeffries] at the Ramsey sprint after his win and not many people knew him.

Talking to Gary Johnson this year in Douglas … [can't be replicated elsewhere] as riders kept away from the public' (22, Visitor, Female, 45–59 years). Interestingly, women were over-represented in the group of respondents citing co-presence with competitors as their defining experience of the TT, suggesting a greater preference for personal interaction with racers, rather than emulation of them.

Whether with friends, family or racers, the idea of motorcycling community is an important characteristic of the TT experience. The Turnerian notion of communitas amongst pilgrims includes absence of socio-economic status or hierarchy, idealised communality and pilgrims as post-pilgrimage change agent (Turner and Turner, 1978), and aspects of these or similar characteristics can be identified amongst TT spectators. In the case of the TT races, the time-space of the event is overwhelmingly dominated by men, which gives a sense of homo-social space and potential for fraternity. Although TT spectators are far from a monolithic homogenous group and evidence a variety of masculinities (Terry *et al.*, forthcoming), these men bring strong shared interests and values, including, in the context of the TT festival, the privileging of motorcycle-centred knowledge and skills over mainstream social norms and socio-economic hierarchies. A number of female spectators expressed similar values.

However, the demographics of TT spectators showed them largely to be made up of middle age, middle class men with disposable income (a finding echoed in a USA study of motorcyclists (Way *et al.,* 2010)); and the attention attracted by high-specification motorcycles, leathers, helmet cameras and other equipment suggest that many TT spectators remain in thrall to an economically-determined hierarchy of material culture associated with motorcycling. Self-identification with the TT event is further materialised and communicated through the purchase of TT-branded merchandise. Like all souvenirs, these items can be likened to tokens of pilgrimage, which identified one as an authentic pilgrim and were a means of materialising both memory of and elements of actual experience in pilgrimage (Shackley, 2006; Coleman and Elsner, 1995; Maddrell and della Dora, 2013). Items such as caps, badges and T-shirts, which are worn on the person (see Figure 5.2) represent a particular embodiment of branded-affiliation with the event. Some souvenirs may even be identified as the secular equivalent of relics. This could include a programme signed by an admired competitor, or, in a parallel with blessed water or soil from a pilgrimage shrine, part of the fabric of the course such as a segment of tarmac from TT landmark Ballaugh Bridge commodified for the TT enthusiast market (see Figure 5.3). More prosaically, TT programmes and other publications are pored over as a source of authoritative knowledge and insight which can be shared with others at the event and taken home as a token of experience for those who were unable to attend.

Interestingly, the Sturgis motorcycle rally held annually in South Dakota, USA, uses very similar rhetoric to that identified in relation to the TT on its website. Sturgis describes its event, now in its seventy fourth year as 'magic', offering friendship, traditional hospitality, fabulous landscape and 'Main Street Madness', as well as opportunities to buy merchandise 'like being a kid again in a candy

Figure 5.2   Branded TT merchandise
Source: Avril Maddrell's own photograph

store' and the 'experience of a lifetime' (Sturgis Motorcycle Rally, n.d.). Given the context of an international shared community of interest facilitated by the internet and international travel, these similarities are not necessarily surprising. However, they are also indicative of an enthusiasm for collective identity and experience expressed and accessed via mass events on the part of a significant number of Western motorcyclists, for whom motorcycling is not simply a mode of transport but an important expression of identity, world view and values. While motorcycling is a minority form of transport in the Western world, and can be read as an expression of alterity by many, those same motorcyclists are drawn to the affirming space of a mass event which reflects their interests and values, provides a sense of larger community, not only accommodates but welcomes their presence, and provides them with a sense of ritualised belonging.

The TT Races and Ritual Practices

In the case of the TT races, for many, the unique opportunity to ride the course is a central part of their TT ritual (Crowther, 2007). Two thirds of those travelling to the TT races with motorcycles will ride the course at some point, with about 50 per cent taking the opportunity to join in the 'Mad Sunday' ride-out when the roads are closed. Thus for many spectators, their most memorable experience of the TT was their own embodied-emotional experience of the course: 'Riding around the

**Figure 5.3   The commodification of relics from the TT course: tarmac
          from Ballaugh Bridge**

Source: Avril Maddrell's own photograph

circuit' (63, Visitor, Female, 45–59 years); '[Riding] fast over the mountain one way' (116, Visitor, Male, 60–74 years). This reflects a sense of co-participation and enhanced respect for the competitors, as expressed by the following respondents: 'You can ride the pack and experience the challenge of the riders' (9, Visitor, Male, 45–59 years); 'You are so much closer to the racers: watching, riding the roads before/after them; it's a unique experience' (58, Visitor, Female, 30–44 years).

Many long-standing TT spectators have other deeply engrained habits and traditions, such as booking the same hotel year on year, having favourite viewing places, meeting the same friends each year on the Island and so on. However, some undertakings are more explicitly encoded *as ritual*. Just as leading competitors can be superstitiously dependent upon rituals and charms (TT champion John McGuiness insists on carrying a lucky penny when he races (McGuiness, 2012)), likewise, spectators can have their own rituals which must be upheld. One group of Belgian motorcyclists recounted how, immediately on arriving off the ferry, 'we all ride to Joey [Dunlop's memorial] statue, we all touch the statue – then the TT begins' [Field Diary, 2010]. This collective ritual act, combines celebration of arrival, the beginning of a special time, and respect for their late TT hero – and perhaps an acknowledgement of the risks associated with motorcycle riding in general and road racing in particular (Dunlop having died racing in Estonia in 2000). Despite their already long journey, for this group of friends, this ritual undertaking of arrival is a required act which takes precedence over the comforts of food, showers or rest. The fact that many TT enthusiasts travel to the Island outside of the race period in order to visit the course and pay homage to the totemic character of the Isle of Man is another indication of the symbolic significance of the TT for road race enthusiasts.

Memorials are micro-sites of homage and ritual practice, which can be attributed with sacred qualities. This is typically the case if associated with bodily remains, articulated explicitly on headstones which read 'sacred to the memory of' (Maddrell, 2009, 2011), or the place of death, especially in cases of untimely or violent death (Petersson, 2010). The symbolic qualities of memorials can also be powerful, notably where the memorial is understood to stand for the person's identity, persona, values and achievements, and/or when it is treated as threshold space of presence and communication between the living and the dead (Maddrell, 2013), as illustrated by Pre's Rock above (Wojcik, 2008). The challenges of the TT course result in regular competitor fatalities and as a consequence the course (and cemetery immediately opposite the TT Grandstand) is home to numerous memorials (see Bradford, 2008 and Figure 5.4). However, less than 50 per cent of spectators had ever visited a TT memorial, with the most commonly visited one (which less than 25 per cent of spectators had visited) being the prominent Joey Dunlop memorial located outside the Motorcycle Museum on the mountain section of the course. This relative lack of active engagement with TT memorials, compared to Pre's Stone for example, is borne out by the paucity of votives, flowers and so on at the memorials, and respondents' negotiation of the inherent risks associated with road racing. The degree of risk taken and skill required by

Figure 5.4   TT memorial wall
Source: Avril Maddrell's own photograph

competitors, brings mortality to the fore and engenders reverence on the part of spectators. This adds to the sacred quality of the event – that competitors will risk their lives in the pursuit of success – but spectators were nonetheless generally reluctant to actively visit or engage with memorials as material embodiment of those risks. A few reported systematically visiting 'all' the TT memorials as a sacred journey undertaken in acknowledgment of the ultimate cost paid by the deceased competitors: 'All, out of respect for all TT riders' (15, Visitor, Male, 45–59 years), but others were more interested in the memorials as heritage, part of the narrative of the event: 'All, I want to see them [*sic.*], know what happened, and to be able to tell other German speakers' (32, International Visitor, Male, 45–59 years). However, the more pragmatic approach of one resident captured the view shared by many spectators: 'Will take friends/others there, but choose not to dwell on them' (83, Resident, Male, 45–59 years). Just as explicit religiosity is marginalised within the TT culture of the TT festival, so too are the sacred dead, who, although frequently revered, may distract from, if not 'haunt', the immediate event were they acknowledged more explicitly by both spectators and competitors.

Conclusion

As evident in other chapters in this volume, the contemporary definition of the 'sacred' is broadly inclusive of spaces, objects and practices which are highly valued. Current definitions of pilgrimage and sacred journeys are likewise flexible. In the spirit of these broad understandings, certain aspects of the TT races and visitor experience of them have been shown to be discursively framed as 'sacred', as an expression of the poetics of enchantment with the TT. Attending the TT is clearly not the equivalent of a religious obligation which is believed to influence one's relationship to the divine or one's status in the afterlife, but the races themselves are represented and marketed via the rhetoric of the sacred, and many spectators attribute sacred qualities to the place, practice and people associated with the TT races. For some, this reflects their personal politics of motorcycling culture in which the temporary conglomeration and privileging of motorcyclists that occurs at the TT represents an ideal world.

Many motorcycle and road racing enthusiasts from the British Isles, Europe and beyond expressed a strong sense that they should attend the races 'at least once in a lifetime' and that doing so is the fulfilment of a lifetime's ambition for those further afield and an annual commitment for many closer to home, designating the TT as a 'sacred' place and experience. These views were expressed by both male and female visitors to the TT races. While it is not surprising that Manx residents did not refer to the TT as pilgrimage, it is notable that none referred to the races or course as in any sense 'sacred', although many revere top competitors and have a deep commitment to the event.

This research has highlighted four key aspects to spectator experience of the TT which can be linked to ideas of pilgrimage and sacred journey. Firstly, the un-prompted self-definition of attendance as pilgrimage; secondly, the long term commitment to attendance on the part of many spectators, which in numerous cases is manifest in annual journeys to the races; thirdly, evidence of rituals and other practices specifically associated with the particular liminoid time-space of the festival; and fourthly, aspects of the 'communitas' associated with pilgrimage, as a result of a combination of enrichment through co-presence and an undermining of usual social hierarchies through the privileging of motorcycling and road racing cultures. A strong sense of being somewhere 'set apart', the biker community, and being outside normal social hierarchies and rules, all contribute to these liminoid qualities of visitors' experience of TT. Annual returnees appear to use the TT races as an anchor or mooring (Hannam *et al.*, 2006) for their identities as motorcycle and motorsport enthusiasts, a source of renewed enthusiasm for the sport and its associated culture. Whilst a number of these observations relate to wider debates about the overlapping nature of pilgrimage and tourism (see the introductory chapter, this volume), if 'pilgrimage is a journey undertaken by a person in quest of a place or a state that he or she believes to embody a valued ideal' (Morinis, 1992: 4), then many supporters travelling to the TT races can be identified, and indeed do classify themselves, as (secular) pilgrims. It may be debatable whether TT enthusiasts are *changed* by their experience of attending the races, as per the models of idealised pilgrimage, but this can also be true of religious pilgrims. Nonetheless, many TT supporters are passionate and steadfast – even 'devout' – in their unswerving commitment to attend the races. Indeed, a full 20 per cent of spectator-respondents reported that only death would stop them from journeying to the TT races.

References

Bradford P., 2008. *Isle of Man TT and MGP Memorials 1907–2007*. Douglas: Paul Bradford.

Coleman S. and Eade J., eds, 2004. *Reframing Pilgrimage. Cultures in Motion*. London: Routledge.

Coleman, S. and Elsner, J., 1995. *Pilgrimage Past and Present: Sacred Travel and Sacred Space in the World Religions*. London: British Museum Press.

Cresswell, T. and Merriman P., 2011. Introduction. Geographies of mobilities. Practices, spaces, subjects. In: T. Cresswell and P. Merriman, eds, 2011. *Geographies of Mobilities: Practices, Spaces, Subjects*. Farnham: Ashgate, 1–18.

Crowther, G., 2007. Embodied experiences of motorcycling at the Isle of Man TT Races [online]. *International Journal of Motorcycle Studies*, Special TT Centenary Edition. Available at: <http://ijms.nova.edu/November2007TT/ IJMS_Artcl.Crowther.html> [Accessed 10 August 2010].

Digance J., 2006. Religious and secular pilgrimage. In: D.J. Timothy and D.H. Olsen, eds, 2006. *Tourism, Religion and Spiritual Journeys*. London: Routledge, 36–48.

Dillon, A., 2012. Olly's pilgrimage. *The Sun*, [online] 3 October. Available at: <http://www.thesun.co.uk/sol/homepage/sport/golf/4568471/Jose-Maria-Olazabal-considers-pilgrimage-to-final-resting-place-of-Seve.html> [Accessed 4 June 2013].

Disko, S., 2007. The image of the 'Tourist Trophy' and British motorcycle racing in the Weimar Republic [online]. *International Journal of Motorcycle Studies*. Available at: <http://ijms.nova.edu/November2007TT/IJMS_Artcl.Disko.html> [Accessed 15 June 2010].

Dubisch, J., 2004. Heartland of America: memory, motion and the (re)construction of history on a motorcycle pilgrimage. In: S. Coleman and J. Eade, eds, 2004. *Reframing Pilgrimage. Cultures in Motion*. London: Routledge, 105–31.

Dubisch, J., 2005. Healing 'the wounds that are not visible'. In: J. Dubisch and M. Winkelman, eds, 2005. *Pilgrimage and Healing*. Tucson: University of Arizona Press, 135–54.

Dubisch, J., 2008. Sites of memory, sites of sorrow: an American veteran's motorcycle pilgrimage. In: P.J. Margry, ed., 2008. *Shrines and Pilgrimage in the Modern World: New Itineraries into the Sacred*. Amsterdam: Amsterdam University Press, 299–317.

Dyas, D., 2004. Medieval patterns of pilgrimage: a mirror for today? In: C. Bartholomew and F. Hughes, eds, 2004. *Explorations in a Christian Theology of Pilgrimage*. Aldershot: Ashgate, 92–109.

Eade, J., 2000. Introduction. In: J. Eade and M.J. Sallnow, eds, 2000. *Contesting the Sacred: Anthropology of Christian Pilgrimage*. London: Routledge. 2nd ed., ix–xxi.

Fern, D., 2003. Motorcycling: TT's fastest rider dies after 160 mph accident. *The Daily Telegraph*, [online] 30 May. Available at: <http://www.telegraph.co.uk/sport/2404984/Motorcycling-TT-races-fastest-rider-dies-after-160mph-accident.html> [Accessed 5 February 2014].

Finnegan, J., 2007. *Emerald Fairways and Foam Flecked Seas: A Golfers Pilgrimage to the Courses of Ireland*. New York: Simon Schuster Paperbacks.

Gemzöe, L., 2012. Big, strong and happy: re-imagining femininity on the Way to Compostela. In: W. Jansen and C. Notermans, eds, 2012. *Gender, Nation and Religion in European Pilgrimage*. Farnham: Ashgate, 37–53.

Hannam, K., Sheller, M., and Urry J., 2006. Editorial. Mobilities, immobilities and moorings. *Mobilities*, 1: 1–22.

Isle of Man TT, n.d., TT Info. Isle of Man TT [online] Available at <www.iomtt.com/TT-Info.aspx> [Accessed 4 January 2014].

Isle of Man Treasury, 2011. *TT Survey 2010*. Douglas: Isle of Man Government.

Jansezian, N., 2010. A pilgrimage for peace and sports. *Travelujah*, [online] 16 April. Available at <http://www.travelujah.com/blogs/entry/A-Pilgrimage-for-Peace-and-Sports> [Accessed 5 June 2013].

Maddrell A., 2009. A place for grief and belief: the Witness Cairn at the Isle of Whithorn, Galloway, Scotland, *Social and Cultural Geography*, 10: 675–93.

Maddrell A., 2011. 'Praying the Keeills'. Rhythm, meaning and experience on pilgrimage journeys in the Isle of Man, *Landabrefið*, 25: 15–29.

Maddrell A., 2013. Moving and being moved. More-than-walking and talking on pilgrimage walks in the Manx landscape, *Journal of Culture and Religion*, 14 (1): 63–77.

Maddrell A., 2013. Living with the deceased: absence, presence and absence-presence, *Cultural Geographies*, 20 (4): 501–22.

Maddrell A. and V. Della Dora, 2013. Crossing surfaces in search of the Holy: Landscape and liminality in contemporary Christian pilgrimage, *Environment and Planning A*, 45: 1105–26.

McGuiness, J., 2012. *King of the Mountain* [TV programme], ITV, ITV4, June 2012.

Milton, K., 2002. *Loving Nature*, London: Routledge.

Morinis, A., 1992. *Sacred Journeys. The Anthropology of Pilgrimage*. Westport: Greenwood Press.

Petersson, A., 2010. The production of a memorial place: Materialising expressions of grief. In: A. Maddrell and J.D. Sidaway, eds, 2010. *Deathscapes. Spaces for Death, Dying, Mourning and Remembrance*. Farnham: Ashgate, 141–59.

Pinch, P. and Reimer, S., 2012. Moto-mobilities: Geographies of the motorcycle and motorcyclists. *Mobilities*, 7(3): 439–57.

Reader I. and Walter, T., eds, 1993 . *Pilgrimage in Popular Culture*. Basingstoke: Macmillan.

Shackley, M., 2006. Religious retailing at Ireland's national shrine. In: D.J. Timothy and D.H. Olsen, eds, 2006. *Tourism, Religion and Spiritual Journeys*. London: Routledge, 94–103.

Stausberg, M., 2011. *Religion and Tourism: Crossroads, Destinations and Encounters*. London: Routledge.

Sturgis Motorcycle Rally, n.d., Are you ready for the 74th Annual? *Sturgis Motorcycle Rally*, [online]. Available at <http://www.sturgismotorcyclerally. com/info-guide/welcome.php> [Accessed 5 February 2014].

Terry, A., Maddrell, A., Gale, T. and Arlidge, S., forthcoming. Spectators' negotiations of risk, masculinity and performative mobilities at the TT races. *Mobilities*.

The Manx Independent, 2011. What brings them back to the TT year after year? *The Manx Independent*, [online] 3 June. Available at <http://www.iomtoday. co.im/news/isle-of-man-news/what-brings-them-back-to-tt-year-after-year-1-3444024> [Accessed 5 February 2014].

Turner, V. and Turner, E., 1978. *Image and Pilgrimage in Christian Culture*. New York: Columbia University Press.

Vannini, P., and Taggart J., 2012. Doing islandness: a nonrepresentational approach to an island's sense of place. *Cultural Geographies*, 20(2): 225–42.

Walter, T., 1993. War grave pilgrimage, in Reader I. and Walter, T., eds., 1993 . *Pilgrimage in Popular Culture*. Basingstoke: Macmillan, 63–91.

Watterson, J., 2010. Survivors moving forward. *Island Racer*, 70–76.

Way, K.A., Roberts L.J. and Turner L., 2010. Travel and spending characteristics of the mature biker: An exploratory study. In: *Emerging Issues and Trends in Hospitality and Tourism Research*. Las Vegas: Hurrah Hospitality Summit.

Whitley, D., n.d. Australian Sporting Pilgrimages. Available at <http://travel.ninemsn.com.au/holidaytype/sport/655285/australian-sporting-pilgrimages> [Accessed 4 June 2013].

Whitney, C.,1895 [1923]. *A Sporting Pilgrimage: Riding to Hounds, Golf, Rowing, Football, Club and University Athletics*. New York: Harper and Brothers.

Wojcik, D., 2008. Pre's Rock: pilgrimage, ritual, and the runners' traditions at the roadside shrine for Steve Prefontaine. In: P.J. Margry, ed., 2008. *Shrines and Pilgrimage in the Modern World: New Itineraries into the Sacred*. Amsterdam: Amsterdam University Press, 201–12.

Wunderlich, F.M., 2010. The aesthetics of place-temporality in everyday urban space: the case of Fitzroy Square. In. T. Edensor, ed. 2010. *Geographies of Rhythm. Nature, Place, Mobilities and Bodies*. Farnham: Ashgate, 45–56.

SECTION II
Tracing Historical Footprints

Chapter 6

Legend Landscapes: Sacred Mobilities in the 'Legend Trip' Tradition

Rob Irving

Introduction

> Once theophanies are localised, pilgrimages necessarily follow.
>
> (Herbermann, 1913: 85)

Herbermann's pithy epigraph suggests a process through which place, as setting, represents an interface where human conceptions of gods or deities are revealed through tangible experience of divine action. At such sites, these conceptions are transformed from the intuited into a lived experience, which is then passed on to others in the form of a legend. This impels believers to mobilise and visit the place in question in the hope of bearing witness to such a drama. This chapter examines how dual senses of expectancy and receptivity combine to influence and even drive this process.

It is not fatuous to say that place is elemental to pilgrimage, since place is what binds the diversity of the cultural identities so engaged, whose myths are played out in the context of psychophysical landscapes and the places that constitute them. It is the way place mediates body and mind, the physical and psychical, the 'real' and imagined, suffusing nature, the supernatural, and art, and how it gives physical presence to projections of the human psyche, that defines my development of earlier notions of the 'legend landscape' (Peuckert, 1965; Dégh, 2001; Irving, 2014). In this self-reinforcing cycle, memory, imagination, and artfulness together contribute to the shaping of legend landscapes as sites of pilgrimage and mobility where otherworldly events are said to have occurred and spiritual presences (and absences) still dwell.

The three examples I shall discuss here of those who visit a particular place in the hope of participating in its legend drama – religious devotees, New Age mystical tourists, and legend trippers – do not involve the same kinds of sacred journeys … each to their own; disclosure and revelation manifest in different ways to different communities.[1] However, what these journeys have in common for the participants

---

1 For example, ghost hunters are less likely than Catholics and Hindus to report moving statues of the Virgin Mary or a milk-drinking Ganesh; religious devotees tend not

is a sense of phenomenological involvement with otherworldly narratives that are peculiar to particular places. This suggests that hierophany finds its value within a perceptual field composed of a virtual reality generated from sensory input filtered through pre-existing frameworks of knowledge, ideas, and poetic and mimetic associations that are determined by cultural praxis.

Much of my research concerns human relations with art objects that are manufactured by human artists but are not believed to have originated that way; instead, as Gell put it, 'they are thought to be of divine origin or to have mysteriously made themselves' (Gell, 1998: 23). To me, as an artist, their success as artworks gains traction when they are not perceived as art, but develop lives of their own as objects of cognition whose capacity to affect an audience depends not only on the objects' own qualities but also on the reader's cultural habits or expectations (Eco, 2011). I am particularly interested in the ritual performance of something like 'worship' through the sanctification of witness and experience. It is important to recognise how place contributes to this performance: the human fascination with 'places of power' intensifies the reciprocal nature of this kind of engagement, where phenomena are experienced as an intimate exchange, or communion, with the otherly, that is peculiar to that place.

In his poem *Description Without Place* (1945), Wallace Stevens defined this kind of creative receptivity as 'an expectation, a desire ... a little different from reality, the difference that we make in what we see' (Stevens, 1945: 563). In terms of pilgrimage, the potential for revelation and/or phenomenological experience is mobilised by being carried to a place made special by myth and legend, and thus into a charged perceptual field. This chapter addresses mobilities less as an analysis of the movement of people – although it is myths and legends about a place that impel people to visit it – than as the traffic of ideas, behaviours, feelings and experience, their association with place, and their cultural circulation and informational mediation as *narrative performance*. In such circumstances, the power of myth-based discourses and practices can be tracked in terms of their value as network capital (Sheller, 2011) within a sacred economy (Morgan, 2012). But first I should explain exactly what I mean by myth and legend in this context – how they are different, and how this difference drives people to visit particular sites in the desire for such encounters.

Myth and Legend

'Myth' and 'legend' are often treated as synonymous, but folklorists recognise a crucial distinction between them. They operate interdependently to inform performative responses to sacred places, but their relationship also acts to ensure the continuity of myth as it is fed back into society as contemporary folklore.

---

to be fascinated by crop circles, and as far as crop circle pilgrims are concerned they are participating in what they believe to be scientific disclosure, not religion.

Humans are inveterate mythmakers. Myths are accumulations of stories, whose meanings are not expressed either abstractly or as fact but are latent, and cannot be told in any way other than by story. Here I address a type of myth characterised by its setting in some kind of hidden otherworld, populated by entities that stimulate fear, or comfort, or a tensile combination of both – evoking a general sense of the numinous. Morgan's definition of numinosity as 'a quality of mysterious and awesome power, other than man (*sic.*) and yet related to him (*sic.*), which is believed to reside in certain objects of experience' (Morgan, 2012: *xiv*, quoting Peter Berger, 1969: 25) is consistent with the idea that a mysterious presence exists in relation to specific settings and objects, which thus offer the possibility of interaction with this otherly power or entity.

Myth is the product of a peculiar characteristic of the human mind: its ability to deal in ideas and experiences that challenge rational explanation because actual instances of the mythical object cannot be isolated. It is important to recognise when considering myths as objects of belief that they should not be understood as 'lies', or necessarily fictions, but rather as mental images existing within networks of ideas that suggest particular ways of interpreting the world and shaping human understanding of it. As such, myths give form to communal experience, representing shared fears or desires that make up a collective ontological imaginary; and this lies dormant until circumstances bring these visions into the open. To rationalists inclined to see reality in purely material terms, these ideas are easily tested and disregarded. However, on another level they are an invitation to proceed beyond prescribed limits of understanding in search of deeper meanings – for, as Bachelard observed, these are 'problems posed by poetic imagination, [where] the cultural past doesn't count' (Bachelard, 1958 [1994]: *xv*). Like art and play, myth is unfettered by reason and logic. For folklorists, myths are neither *necessarily* true nor false[2] – rather, they are stories that may be distinguished by their ability to take teller and listener into realms that exist at least partly beyond everyday experience. These provide a structured frame of reference in which to place the unknown, counterbalancing what is already known and revealing potential new knowledge. To the folklorist, the truth or falsity of, say, a particular religious belief is immaterial: the focus (as in my treatment here) is on its dynamics. The myth of 'life after death', for example, is usually themed around familiar elements that would appear to be designed to provide comfort, or evoke fear, for those who take it literally. It is also counterintuitive: the imaginary counters the material evidence that death ends life and our bodies perish. But when this or any myth is coupled with religious ideas, for example, doctrines of redemption and/or karma, it acts to modify behaviour in the living.

It is at this interface of epistemological values that myth and legend become intertwined, as is common in much of heritage tourism (Ramshaw 2014). Whenever

---

2   Gadamer (1980: 92) writes: '*the only good definition of myth is that myth neither requires nor includes any possible verification outside of itself*'. [Emphasis in original] (Pflock and Brookesmith, 2007: 163).

myth is represented or illustrated by an event that is said to have occurred in real time and space – and therefore invites dispute – it becomes legend. To illustrate this using a well-known example: In *The Legend of Sleepy Hollow* (Irving, 1820), the protagonist Ichabod Crane sees a headless horseman, who is the personification of the myth of revenants. That these exist in the flesh, as it were, is the myth; the legend concerns the specific instance that was witnessed at a particular time and place identified in the narrative. Another example is that of Marie-Bernarde Soubirous and her friends' encounter in 1854 with 'a small young lady' who identified herself as the 'Immaculate Conception'. The legend of Lourdes, France, is agentive of the myth of Mary, mother of Jesus, and that of his virgin birth, which had become Catholic doctrine four years earlier, as well as of the site of this encounter as a place of healing. For her role in this, 'Bernadette' (as she is better known) was declared a Saint of the Roman Catholic Church, and Lourdes has become one of the world's busiest pilgrimage sites.

Legend may be passed on by word of mouth, but in the kind of situations I am discussing here it is often acted out or performed as real life, and may involve what performance theorist Richard Schechner calls 'dark play' (Schechner, 1993: 27), in which one person is aware of pretence while participants in the drama it provokes are not. The semiotic implications of this kind of performance apply to things as well as people. For instance, a legendary object, such as a hoof print at Sleepy Hollow, not unlike a crop circle, tells a story through the agency of its association with a legendary event, bringing myth to life and into legend. If the Blessed Virgin had left a physical artefact behind at Lourdes it would now be a sacred relic; her physical legacy is the legendary healing power of the waters. Indeed, this transformation of the spiritual into the physical defines the criteria used by the Catholic Church to determine the legitimacy of Marian apparitions (*Catholic News Agency*, 2008).

To recapitulate: while myth lies across the threshold of everyday reality and otherly realms of the imagination, legend occupies another liminal space, between fact and fiction, as expressed conceptually within certain social environments. Legend pulls myth into the everyday world of actuality. Myths may be a comfort, legends are not; legends are designed to test the permeability of the *status quo*, and as such may represent a contradiction of conventional values. The relevance of place here is that it emplaces the human body, giving form to ontological and epistemological tensions, and the literal and/or metaphorical interpretations of phenomena. When we associate certain places with a particular legend, these become the embodiment of the legend – for example, a haunted house embodies, even *personifies,* the idea that ghosts inhabit places. What makes this so powerful is that it is not the product of thought alone, but is also *embodied as* felt experience, and enables us to situate ourselves within the mythical environment. Douglas's observation that 'a marked off place alerts a special kind of expectancy, just as "Once upon a time" creates a mood receptive to fantastic tales' (Douglas, 1966 [2002]: 78) is pertinent to this argument.

As Casey (1987: 202) observed, the task of place is to congeal this poetic imaginary into a provisional reality. It is possible to tie these abstractions into specific kinds of experience. Sallnow notes that when revelatory phenomena are associated with specific places they can be viewed as either emanating directly from the matrix of an animated landscape, or as a consecration of the territory *via* apparitions and physical manifestations by 'external, foreign divinities, whose power is *sui generis* but who choose to channel it through their particular sites to particular populations' (Eade and Sallnow, 1991: 6). While earth spirits, and the popular New Age notion of 'earth energies' are examples of the former, the latter would include UFOs (Unidentified Flying Objects, a euphemism for extraterrestrial craft, 'flying saucers' and so on), and apparitions of the Blessed Virgin Mary. However, this is not the case for all apparitions; ghosts may be thought of as haunting, even trapped in, a particular place. A range of other phenomena, such as crop circles and *ignes fatui*, sit in either set, or somewhere in-between, depending on what their experiencers, guided by cultural forces and expectations, think they represent.

My interest here is specifically in the place of mobility in such transactions, and my main focus concerns the Avebury area, Wiltshire, and its attraction for New Age pilgrims. I will examine the performative activity that takes place there in light of older traditions of place-based encounter. But first I will introduce the concept of the 'legend trip', and isolate the similarities between the ostensibly disparate kinds of pilgrimage that the legend trip involves.

The Legend Trip

Ellis (2001) defines the legend trip as a journey to a site where something extraordinary is said to have occurred; it can be located on the continuum of places, events and experiences identified as 'Dark Tourism' (Lenon and Foley, 2000), this in turn being closely linked to heritage tourism (Sharpley and Stone, 2009). The legend narrative usually features a close, often grisly encounter with a mythical entity that is believed to haunt the site. One example is a part-man, part-goat satyr known as the Pope Lick monster, said to live beneath the railway trestle over Pope Lick creek in Louisville, Kentucky, and blamed for a number of deaths at the trestle since its construction (Kleber, 2000). It is important for the structural context of the legend trip that these occurrences are unpredictable and that the legend therefore communicates the implicit suggestion that what happened at the site could happen there again. This adds a vital element of risk, thus eliciting in the legend tripper a sense of expectancy at the prospect of becoming an active *participant* in a continuing social drama. Germane to my argument is a process known to folklorists as ostension, or legend telling by action (Dégh, 2001; Ellis, 2001). The legend trip is similar to what Ellis has termed the 'ostensive ordeal' but where he sees this activity as staged in a traditional sense, involving pretence rather than belief, what I am describing is played out as real. Even when the drama involves dark play, a hoax cannot really be described as such when it is not perceived as false by participants in its performance. From the point of view of truth-conditional semantics, when artifice creates something that is contrary to fact it becomes legend. The extent to which we suspend critical judgment as we play along with cultural legendry is an open question. But it is important to recognise that it is quite human to want to prolong the delight of not knowing, and that there is method in it. Festinger's observation that cognitive dissonance 'may be reduced, or perhaps eliminated completely, by changing one's own opinion so that it corresponds more closely with one's knowledge of what others believe' (Festinger, 1957: 182) is appropriate here. Moreover, when this heightened sense is shared with other participants it provides a basis for *communitas*, where individuals temporarily identify themselves and one another as members of a unified body (Turner, 1969: 96–7; Ellis, 2001: 183; Carse, 2008: 84–6; Turner, E., 2012). In folklore studies the legend trip tends to be portrayed as a rite of passage for adolescent thrill-seekers, whose aim is to return with a spine-tingling story to tell, placing them at the centre of the legend. When such experiences are received and/or passed on as circumstantial reality, this process creates *new* legend and thus rejuvenates the underlying myth of the phenomena it purports to represent.

Within the corpus of modern folklore, the classic legend trip destination is the haunted house. But abandoned factories, remote roads, bridges, churches, uninhabited areas such as woods and moors, battlefields, cemeteries and so on, may evoke the same response. Can a cemetery really be said to be uninhabited? This question burns at the core of the legend trip experience. Place brings influence to bear on how semiotic objects are perceived. By 'semiotic object', following Eco (1979; 2011), I mean a set of cultural properties that is conveyed by a given expression or stimulus. Certain supernatural semiotic objects seem to belong in certain places but would seem ungrammatical, or out of place, in others. To regard the cemetery as inhabited by spirits of those people who are buried there is akin to Douglas's 'special kind of expectancy' associated with a sacred place. This invites an attitude of receptivity, creating ideal conditions for phenomenological encounter, and *reciprocal activity* on the part of recipients, offering people a participatory role in the received notions of how such encounters are played out.

Moving Statues and Animate Images: The Reciprocity of the Pilgrim Gaze

An example of the melding of religious pilgrimage and legend trip occurred in the summer of 1985, when someone passing through Ballinspittle, County Cork, Ireland, noticed movement in a neon-lit roadside statue of the Blessed Virgin Mary. That year, Ireland was experiencing a widespread enthusiasm for Marian visions and similar phenomena. Word of this occurrence spread fast, drawing large crowds to this otherwise unremarkable village. The subtle movement of the statue was subsequently witnessed by thousands of people. Events of this kind can open rifts between doctrinal ecclesiastic attitudes to divine interventions and spontaneous outbreaks of Marian phenomena, which are often treated by the

Catholic Hierarchy, at least initially, as heretical. What is interesting about this example is how it offered pilgrims the opportunity to participate in the legend of the miracle as eyewitnesses to it, and even creators of it in their own performance, possibly by unconscious optical illusion. An eyewitness, former BBC religious affairs correspondent Dr Ted Harrison, offers a clue to this:

> Had I been looking for a miracle, I might have claimed that I'd seen one. My response, however, was more dispassionate. I reviewed in my own mind what I had seen and looked around me at others who, presumably, were seeing the same thing. I noted firstly that at the precise moment I saw the statue move, no one else claimed to have done so. Indeed, as far as I could see, *no two people ever saw movement at exactly the same time as each other. This suggested that the movement was not attributable to the statue, but to the observer.* [author's emphasis] (Harrison, 2011)

These events were reminiscent of another example from 130 years earlier: On an August night in 1879, a vast illumination resembling devotional images attributed to Bernadette's vision at Lourdes appeared outside the village church at Knock, County Mayo. These images were accompanied by a depiction of Joseph and another of John the Evangelist. The figures were variously described as 'appearances', 'likenesses', or 'statue-like', which, despite remaining motionless, appeared to be 'alive' (Carpenter, 2011: 106). We have here an example of the inherent tension of legend, keeping reality and illusion, art (divinely inspired) and artifice (man-made simulacra of the divine) suspended together in mutual animation. In this case, it pits those who believe the literality of the phenomenon observed against the idea that someone – prankster? priest? – used a magic lantern to create an effect that at that time was certain to impress country folk – twin truths, like boxing hares, as Kathleen Jamie (2012: 37) put it.

Subsequent research (Hynes, 2008; Carpenter, 2011) makes a sound case for the latter theory. In 1880, journalist and photographer T. O'Connor identified from eyewitness reports typical characteristics of projected illumination, such as the illumination's circular shape and its sharp definition from the surrounding darkness. Marina Warner points out that 'the images projected by phantasmagorias swelled and shrunk [*sic.*], as well as shifting with tricks of the lights, and so created an illusion that they possessed the quality of conscious life: animation' (Warner, 2006: 148; cited by Carpenter, 2011).

The visual dynamic that underpins this kind of interactive encounter can be compared with the Vedic concept of *darśan,* which describes not only seeing a manifestation of a sacred or divine presence, but also the simultaneous sense of *being seen by it* (Eck, 1981; Gell, 1998: 117–18). This relates to ancient traditions where the eye is an organ of interpersonal transactions, and sight is perceived as a form of touch: To catch sight of the object of one's own gaze and to be seen and singled out by it is to experience one-to-one contact (Eck, 1981 [1998]: 9).

Animacy originates in this ocular exchange, argues Gell (1998: 118), because of the human capacity for inferring intentionality on the part of images/objects that mirror the perceptions and expectations of the spectator. Just as resonating sounds and echoes can be said to converse with each other, a union is achieved when the object mirrors and is assumed to *reciprocate* the gaze of the spectator. As this oscillating cycle of feedback between seer and the seen shortens it intensifies the visual hierophany, creating 'a more psychological, non-linguistic view of religious vertigo' (Gell, 1999: 17).

This process enables spectators or devotees to see themselves as the deity or object of their gaze sees them, through its eyes, as it were. When the object is environmental – for example, localised weather conditions, or the constituent parts (hills, mounds, trees) that make up 'place' – the suggestion of immanence is consistent with Sallnow's remarks, quoted earlier, about how certain sites seem to exude an aura that emanates directly from an animated landscape, reflecting a sense of 'witnessing as agency' (Gell, 1998: 116) where objects of belief are expected by pilgrims to manifest in terms of observable phenomena.

My fascination with this activity as part of the seepage and traffic between semiotic objects of modern folklore, religious experience, and conceptual/visual art, is similarly in keeping with Vedic tradition, which is open and pragmatic in recognising that it is the human presence that calls the presence of the holy into its own image in order to facilitate communication and to receive *darśan.*

My point is that art (that is, performativity between people and things as social agents) can be conceived of in similar terms as acts of 'making special', which Dissanayake (1992 [1995]) identifies as an elemental aspect of human existence. As with any ritual activity, ostension and its reception involves surrender on the part of the recipient, a giving oneself over to something. This kind of transaction is a creative, collaborative process; the relationship between expectation, revelation, and *self-deception* is finely balanced, and it is in itself a form of collusion in the sense that both seer *and the seen* are agentive.

As myth and legend live outside accepted scientific or materialist values, the idea of 'inanimate' objects as social agents presents problems for any inquirer who is not willing to enter into a transitory state of double-mindedness – typified by the paradoxical nature of the burlesque origins of parody and other 'para' concepts (notably 'paranormal') as belonging *inside* a system that includes things *and* their opposites. Such 'pararealms' offer ideal conditions in which to exploit the human proclivity to use art and artifice as a lens to see beyond everyday realities. They are the natural habitat of Trickster-like activity as a mode of artistic performance that catalyses transformation. Contrary to Western convention the Trickster's intention is to create more, not fewer, layers of complexity than it presumes to resolve, leading to a plurality of possible outcomes; rather than bowing to the principles of Occam's Razor, it acts to subvert them (Irving, 2014). In recent years, this kind of activity has affected perceptions of Avebury and other prehistoric landscapes in terms of contemporary culture, and New Age mobilities of belief and pilgrimage in particular.

Avebury's Legend Landscape

The Avebury complex consists of six key prehistoric megalithic and earthen monuments within an area of 8.7 square miles, around 20 miles north of Stonehenge in Wiltshire, England. This area is designated a World Heritage Site because of the importance of its Neolithic remnants and their placement within a 'ritual landscape' (Robb, 1998; Pryor, 2004: 244).[3] Where once cereal cultivation shared space with ritual activities around cycles of life and death, today it competes with an emerging tourism industry, creating palimpsestic tensions between continuous agricultural utility and a mystical intimacy with ancestral ruins. Over recent decades, the area has become a site of heritage tourism, attracting an annual influx of more than 250,000 visitors (*Which?* 2013). On a busy day, it is possible to observe, through various quasi-religious practices such as dowsing, or using sound – drums, gongs, 'singing' bowls, 'wind harps' and so on – to resonate with place, a communion with ancestral spirits and a cultural overlap with legend tripping. As such, these and similar activities present interesting insights into contemporary sacred journeying in both physical and psychic senses.

Figure 6.1   Every summer for the past 25 years crop circles have appeared in view of Silbury Hill, the largest man-made prehistoric mound in Europe

Source: Author's own photograph

3   While some archaeologists consider the term 'ritual landscape' to be problematic, I am interested in the concept as it is accommodated into New Age beliefs, so I choose to follow Pryor and use it here.

To geographers, historians, and archaeologists, Avebury's ancient topography represents an ensemble of artefacts that make possible the retrieval of physical trace information. This provides a basis for conjecture regarding the conditions of past human inhabitation. The significance and efficacy of modern myth and legend in shaping our perceptions of the material world becomes apparent when these disciplines converge with others concerned with otherworldly social relations. To assume that the escalation of heritage tourism to Avebury is due to the popularity of mainstream archaeology – the *Time Team* effect – would be to ignore its coincidence with the emergence in Britain of a New Age movement, and popular fascination with its foundational myths about UFOs, Earth Mysteries (ley lines and so on), and, more recently but not least, crop circles.

To New Agers, one of the attractions of Avebury, and prehistory as a whole, is that it represents a *tabula rasa* on which to inscribe new histories. But amid the din of conjecture scholars generally agree that Britain's prehistoric monuments were intended by their makers to embody psycho-social internalisations of myths concerning ancestral memory and the celestial forces that create and control life (Darvill and Wainwright, 2005; Pryor, 2010; Parker-Pearson, 2012). That is to say, modifications to the landscape were a way of using the physical to express the intangible. As such, they may be regarded as conceptual artworks, rather than merely artefacts of only antiquarian interest, and should be understood as possessing liminal or otherworldly characteristics that make sites, and sights – for many sites are placed to be seen from afar – special in terms of the social identities played out here. The effect/affect of these interventions is evident today in the ritual transactions and social participation they attract, particularly, as noted, as sacred sites of New Age mystical tourism, replete with immanent vitality in the form of 'earth energies' and intervisiblity.

Cosgrove's observation that landscape and myth represent distinct but articulated signifying systems through which social relations among individuals and groups, and human relations with the physical world, are reproduced and represented [Cosgrove, 1993: 281, citing Short, 1991, and Duncan, 1990] is pertinent here. This ritual landscape is the kind of place-world Casey identifies as 'thin', revealing 'a lability of place that corresponds to a fickle self who seeks to be entertained' (Casey, 2001: 684–85). Mysterious artefacts become transformative, catalytic, absorbed into ritual acts of re-membering as legend, at which point the materiality of this ancient place becomes thick again with meaning. Thus, the psychological impacts of artworks and their mythic undercurrents affect the way sites are perceived and experienced through social processes of make-believe. In this sense, the crop circles that attract so many pilgrims to the Avebury area act in the same way as ancient interventions in the landscape as a way of using the physical to express the intangible, and to physically *situate* visitors in the landscape, and therefore may also be regarded as conceptual artworks, albeit temporary ones.

While pilgrimage tends to be associated with religious activity, it also relates to nostalgia; as such, it is mobility across time. This, and particularly the goings-on in Avebury's ritual landscape, can be related to Smith's model of the 'counter-tourists' who reject received wisdom and 'subject the sites they visit, heritage or otherwise, to their own associations, stories and reconstructions, [and] are not passive, but agentive' (Smith, 2012: 207).

Crop Circles

As far as the physical reality of the patterned areas of flattened plants we know as 'crop circles' is concerned, there is no mystery to their appearance; people make them, usually illegally, under cover of darkness. Yet, crop circle *aficionados* (known as 'croppies') go to great lengths to preserve a sense of elemental mystery in both the circles' appearance and their efficacy. This determination reflects one of the fundamental impetuses of the New Age: as a movement of renewal, it expresses a conceptual imperative that privileges conjecture over the kind of problem-solving that deploys scientific knowledge and methods. Among croppies, the notion that crop circles present a scientific problem persists in the face of an embarrassment of evidence to the contrary. In what is surely proof of the performative, even theatrical aspects of this activity, croppies and scientifically-oriented skeptics have long engaged in disputes, while (perhaps more significantly) ideologically distinct bands of croppies tangle continually in bitter internecine conflict. Meanwhile, a spectrum of the crop circle community from circle-makers to website administrators to conference organisers, tour operators, experts, souvenir sellers, and even visitors to crop circles, are complicit in maintaining the 'mystery'. Beyond this subculture, the legend is nourished by print and electronic media that extend worldwide.

In the 1950s, Jung identified the tendency of the media to affirm these kinds of modern myth as a reflection of the public appetite for mystery and/or quasi-religious iconography (Jung, 1959 [1987]: *x*). In doing so, he raised an interesting question about our hankering for strange phenomena. This cannot be answered satisfactorily by simply reducing the problem to its material constituents and containing it within what is already known; to do so would ignore the crucial element of desirability. Rather, the performativity surrounding such phenomena may be an attempt to escape the confines of rationalism.

As its name suggests, the New Age can be viewed in terms of a re-awakening, but it is more accurately characterised as the re-imagining of an ancient dreamtime. In being re-imagined, so the world is re-made. Paradoxically, this approach is attendant upon a need for the authority of contemporary science to validate its claims to an alternative truth. The boxing hares of art and self-deception are very active in crop circle science, or 'cereology', reflecting New Age tendencies to embrace pseudoscience, favouring ideas that present an anti-structural mirror image of scientific orthodoxy. Dowsing is a good example as it expresses unobservable objects of belief – such as subterranean energies – in terms of observable phenomena, authenticating (and authorising) their existence. The overarching aim here, like all ritual practice, is transformation: to dissolve scientific and rational boundaries, returning us to a state of affairs where there is no distinction between knowledge and spiritual belief.[4] Much of this alternative gnosis is projected in the form of conjecture about the prehistoric past and its sites and landscapes. The image of Avebury and its environs as a playframe typifies the Aristotelian notion of a place that is both receptive and preservative of the past. It offers liminal conditions for pilgrims to experience a shared imaginary, thus restructuring New Age sensibilities in powerfully aesthetic terms (Morgan, 2010: 7).

Crop circles function on multiple levels, not only as 'gifts from the gods', or even *sui generis* phenomena that evince occult power, but also as sites of ritual performance. As such, on one level they are catalysts of conviviality, places where croppies meet and form social bonds. On another, more esoteric level, such sites represent still points in a turning cosmos; they are temporary temples of contemplation, serving as sensorial 'sites of encounter' with their physical surroundings and the occult presences that are thought to reside there. Here, 'field' becomes both the literal and metaphoric interface by which the human body can make contact with these presences. I witnessed an example of this while visiting a crop circle overlooking Silbury Hill, where I overheard, and documented, two people communing with the True Circlemakers, whom they named as the Star

Figure 6.2   Crop circles serve as 'sites of encounter'; here, overlooking Silbury Hill, as described in the text, two pilgrims communicate with the otherly entities they believe created the pattern

Source: Author's own photograph

---

4 The 'post' in post-rationalism, as with post-modernism (Latour, 1993; Locher, 1999), is actually *pre-modern* in its leanings. To Latour, post-moderns feel that they come 'after' the moderns, but with the disagreeable sentiment that there is no more 'after'. 'No future: what remains?' What remains is the vision of a return to pre-modernity.

Councils of Light: 'Did you make this circle?' one asked her moving pendulum. 'Yes' came the reply, which she passed on to her friend and to anyone in earshot. For this person, as well as acting as a kind of telephone, the crop circle, thus verified as authentic, also authenticated the sanctity of its setting within this peculiar landscape. *Murtis* in Hindu temples and shrines, as embodied expressions of divine spirit, perform a similar role, mediating communication, providing the technological means to connect with particular deities.

Crop circles also have extended selves as quasi-religious graphical motifs that circulate beyond their physical location, and therefore take on a mobility of their own in the form of distributed agency. Photographs of the 'signs' *in situ* affirm a mystical association with particular landscapes; souvenir images that can be taken away or disseminated as tokens (postcards, calendars and so on) are an essential part of the pilgrimage experience (Freedberg, 1989: 100; Reader, 2011). Moreover, the motifs are interpreted as symbols from mystical alchemy, rune languages, mandalas, and sacred geometry, and may thus be considered as devotional art, and of the highest quality (Schnabel and Irving, 1992). These iconic motifs are absorbed into a mystical framework, and channelled through contemporary culture in a continuous feedback loop. Something is being communicated: Is this evidence of 'the gods' displaying an ability to interact with humans or a conversation we are having with ourselves? In creating this ambiguity, human artists are not merely catering to a disarticulated artworld view – there is little that is transformative in that – but are integral to a social process of creative stimulation by inviting the rest of us to wonder if what they produce is humanly possible, and, so it follows, 'if they are not made by "us", then by whom or what?' At least, that was the original idea; as with all legend this one mutates according to shifting conditions.

In considering this particular artistic intervention, it is important to recognise that its creative scope extends to include the discourse the crop circles evoke. In 35 years or so, crop circles have evolved from being perceived as physical imprints of flying saucers to having an innate power to act upon us either directly or remotely, as images. Like moon rock (Pearson and Shanks, 2001: 95), 'genuine' (that is, supposedly non-man-made) crop circles possess an aura (Benjamin, 1935; Taussig, 1987; Pearson and Shanks, 2001) that denotes otherness. This aura[5] and its efficacy are predicated on the idea that the works are non-human in origin, which is contingent on their sudden occurrence and the hiddenness of their maker/s. Thus, the effect of the artwork on spectators is fundamentally conditioned by the collective, subjective notion of the processes which gave rise to it.[6] Continuing

---

5   Aura refers here to the vitality of an object's cultural relevance. Benjamin defines this as where 'true to its nature', the object remains 'distant, however close it may be'. (Benjamin, 1935)

6   One example of this occurred in 1993, when iron filings deposited in a crop circle in the rain, and which had subsequently congealed into a rusty glaze, were identified as evidence of high heat (Burke and Levengood, 1995), supporting the authors' pre-conceived notion of how crop circles are created.

Pearson's and Shanks' analogy, however, if the moon rock were revealed to be not from the moon, but from earth (that is, it would be just another piece of rock), its aura vanishes; once exposed as man-made the crop circle reverts to being 'only' flattened wheat. Viewed in these terms, 'genuine' crop circles give presence to absence, and therein lie their power as artefacts: as they imply the invisibility in the here and now of their makers, they foster a *climate* of Barthesean discourse about authorship and intent. Taken individually, crop circles are effective because they are *affective;* once identified as man-made art, they are unable to sustain the effects/affects ascribed to the 'genuine' non-man-made model, even when they are physically identical. To reiterate, *these artworks are made to be consumed by people who do not regard them as art.* As soon as they are seen as art, the magic contained within them, or within us, is diminished.

Figure 6.3   A crop circle is used as a place to perform spontaneous rituals and blessings, often involving sound/resonance
Source: Author's own photograph

Therefore, in understanding the crop circle pilgrim experience it is important to recognise the value of keeping alive the aura and efficacy of crop circles as ritual objects – even as their mystery is assumed to be (and promoted as) a scientific problem (falsely suggesting an intent on the part of 'researchers' to solve the mystery). This pretence is integral to their performance. As their auric 'distance' is believed to evade normal scientific evaluation they invite novel approaches to solutions to the problems they pose, opening the discourse to include the lay audience and not just 'experts'. The performative potential of this interactive dynamic is exemplified by this promise-charged advertisement for an 'adventure of a lifetime' summer crop circle tour in July 2013:

Identify the Energy

This year's mission: identify the source of the mysterious energies that emanate from Balls of Light seen in and around the crop circles. Despite our best efforts, we still haven't pinpointed exactly where this energy comes from. In the past, it has destroyed batteries and rendered cameras useless, some permanently. And we should warn you; this tour is not for everyone. We're only looking for people that want an adventuresome, cutting-edge, mystically-filled tour on the edge of science. So if you are prepared to experience and explore new ideas and new realities, this tour is for you!

Unpredictable Effects

While we will take every step to insure your safety, we still don't completely understand crop circles and don't know what effect they'll have on you. (Please do not sign up for the tour if you have a pace-maker.) In the past we've experienced battery failure, camera malfunctions, and laptop anomalies. Each crop circle is different so understand that you go into these circles knowing and accepting that anything can happen. To date, crop circles have been associated with space and time distortions, strange lights, and UFOs. For these reasons, we'll ask you to sign a waiver of liability form before the tour starts. (Crop Circle Tours, 2014)

As *murti,* crop circles perform as channels into magical thought, and through this, purported new science. At $6,000 per head, plus travel, it is unsurprising that participants strive to get full value by justifying their belief in the adventuresome nature of the field trip by re/enacting, and thus realising, its fantastic possibilities. This makes possible, even *probable,* the misinterpretation of naturally occurring events as experience of an existing legend, for example, thistledown seen through the telescopic lens of a video camera as 'balls of light' – even 'observing entities' because of their close proximity to a crop circle.[7] This tour, and others like it, are synonymous with the traditional legend trip described earlier; thrill-seekers return thrilled as their heightened sense of expectation manifests as anomaly in some form or another. And so myth lives on, rejuvenated by legend, by an accretive and continuous system of adaptation and mutation, according to shifting cultural tastes. Such dramas may be understood in terms of Morgan's notion of the sacred economy, which relies on a different system of credibility from secular economies,

---

7   The fact that these 'balls of light' are only visible to the person looking through the camera is said to contribute to the mystery, intensifying the sense of the sacred revealing itself to particular groups or individuals (Irving, 2014).

'invok[ing] different authorities to bolster the terms of exchange' (Morgan, 2012: 85). Within the locus of religious experience, it may seem ludicrous to compare, for example, a moving statue or vision of the Virgin Mary, with cleverly flattened wheat, but it is what is *brought* to this kind of transaction that defines the object's aura and power to effect its environment, and visitors. The site-specificity of the legend is determined by belief orientation on the part of the actors in its performance, to which place itself makes an essential contribution.

Morgan's notion also relates to the commerce, and competition that exists in the monetisation of crop circles through the sale of aerial photographs, DVDs, books, calendars, postcards and so on. As previously mentioned, souvenirs are integral to the pilgrimage experience. These things, too, become mobile – something to take back, to trigger memories and reconnect the pilgrim to numinous affect, adding to the authority and *authenticity* of the sacred journey. Similarly, relics taken from crop circles, including hallowed earth, plants, water (bottles buried or blessed in a crop circle) and stones, also become mobile. Conversely, in a manner analogous to the exuviae used in ritual magic, things left behind are also a form of interaction in which religious experience is not only mediated by objects but also by place: the idea that a personal object remains in a specific location can be highly evocative, suggesting efficacy on the part of the pilgrim as a participant in the spiritual exchange. Examples of this at Avebury include effigies, crystals, flowers, sage and so on, placed under standing stones, or 'clouties' hung in particular trees, such as the willow at Swallowhead Spring close to Silbury Hill.

Conclusions

The aura of the crop circle, the movement of the holy statue or image, and things that go bump in a haunted house, sensed as distant communication from mythical otherworlds, may be contrived by a delicate balance of complicity and collusion, psychically insulated from everyday reality within a perceptual field. Framed in terms of a sacred economy, expectancy is important because this kind of activity invites a peculiar attitude of receptivity, and *reciprocity* on the part of recipients. The kind of modern myths I have discussed here survive by the movement of their localisation not only in relation to place but also context. This is grass roots legend politics, and its cultural mutability is itself a form of mobility.

Peirce's observation of how semiosis occurs *ad infinitum* throughout communicative processes whenever truth and fiction, fantasy and reality come into contact echoes the semiotic dynamic described in Turner and Turner's analyses of pilgrimage, and the *performance of* and *responses to* the semiotic or social object as a 'vehicle' (Peirce n.d., cited in Hartshorne and Weiss, 1931; Turner and Turner, 1978: 143) carrier of story. This suggests a kind of mobility that is not normally associated with pilgrimage but which, I would argue, applies to the folklore of Christian and other established religious cultures as much as it does to New Age and paranormal legendry. Metaphor, movement and flow are integral to mythical intervention, and to the ways in which Trickster-like behaviour mediates between opposites, activity that is consistent with early Greek notions of flux and poïesis, articulating relations between make-believe (*mythos*) and 'rational' *logos*, in an effort to open new ontological vistas.

Parallels may be found in Indian storytelling tradition. Legends are not intended to conform to ordinary truth-values. Indeed, in both Indian storytelling and legend-charged environments, the more fanciful the telling and retelling of the legend the more 'true' it seems to become (inasmuch as it represents underlying truths, and anxieties, that bind its believers, thus maintaining the gap between everyday truths and 'the mythical realm of the imagination [where] what is important is the truth as we see it' (Stevens, 1951: 147). This may also be likened to the conditions of a séance, where a non-event would be a rare event indeed. The same inversions can be observed in both religious pilgrimage and legend trip, where legendeers enter legend environments – a shrine; a haunted house; a prehistoric landscape – in anticipation of encountering legendary semiotic objects. Hearing the stories that ensue, one senses an underlying challenge that is also evident in the Indian tradition: 'How far can I take my tale before it is disbelieved?' In legend environments, the most outlandish stories are often the most socially binding, while disbelief can mean alienation. As such, the art of legend telling – orally or by action – may be regarded as a systematic filtering of disbelief, leaving a sediment of the willing to seek, and believe in, that 'third thing' (Eco, 1989: 49), which is indigenous to the gap, connecting disparate realities. Hence, even the most 'unbelievable' evidence is embraced at face value; its sheer *irrationality* acts in its favour. As Tertullian reputedly said: *Credo quia absurdum est*. Inside these looking-glass environments, facts no longer require proof and explanation because 'reason' is no longer normative and self-explanatory. Such legend landscapes are phenomenological dream-worlds that generate a note of indeterminacy that sings across thresholds.

References

Bachelard, G., 1958 [1994]. *The Poetics of Space*. Boston: Beacon Press.

Benjamin, W., 1935. *The Work of Art in the Age of Mechanical Reproduction*. Available at <http://design.wishiewashie.com/HT5/WalterBenjaminTheWorkofArt.pdf> [Accessed 20 January 2014].

Berger, P.L., 1969. *The Sacred Canopy: Elements of a Sociological Theory of Religion*. Garden City NY: Anchor Books.

Burke, J. and Levengood, W.C., 1995. Semi-molten meteoric iron associated with a crop formation. *Journal of Scientific Exploration*, 9(2): 191–9.

Carpenter, P., 2011. Mimesis, memory and the magic lantern: What did the Knock witnesses see? *New Hibernia Review*, 15(2): 102–20.

Carse, J.P., 2008. *The Religious Case Against* Belief. New York: The Penguin Press.

Casey, E.S., 1987. *Remembering: A Phenomenological Study*. Bloomington, Indiana: Indiana University Press.

Casey, E., 2001. Between geography and philosophy: What does it mean to be in the place-world? *Annals of the Association of American Geographers*, 91: 683–93.

Catholic News Agency (CNA), 2008. Expert explains Church's criteria for confirming Marian apparitions *Catholic News Agency* [online] 8 May 2008. Available at: <http://www.catholicnewsagency.com/news/expert_explains_ churchs_criteria_for_confirming_marian_apparitions/> [Accessed 3 June 2013].

Cosgrove, D., 1993. Landscapes and myths, gods and humans. In B. Bender, ed., 1993. *Landscape: Politics and Perspectives*. Providence and Oxford: Berg, 282–305.

Crop Circle Tours, 2014. Crop Circle Tour 2014 – Experience the Resonance and Subtle Energies of Wiltshire in Britain. [online]. Available at: <http://www. cropcircletours.com/croptour> [Accessed 20 January 2014].

Darvill, T. and Wainwright, G., 2005. Beyond Stonehenge: Carn Menyn and the Bluestones. *British Archaeology*, 83(July/August): 45–52.

Dégh, L., 2001. *Legend and Belief: Dialectics of a Folklore Genre*. Bloomington and Indianapolis: Indiana University Press.

Dissanayake, E., 1992 [1995]. *Homo Aestheticus: Where Art Comes From and Why*. 2nd ed. Seattle and London: University of Washington Press.

Douglas, M., 1966 [2002]. *Purity and Danger*. London and New York: Routledge.

Eade, J. and Sallnow, M., 1991. *Contesting the Sacred: An Anthropology of Christian Pilgrimage*. London and New York: Routledge.

Eck, D., 1998. *Darshan: Seeing the Divine Image in India*. New York: Columbia University Press.

Eco, U., 1979. *A Theory of Semiotics*. Bloomington: Indiana University Press.

Eco, U., 1989. *Foucault's Pendulum*. London: Secker and Warburg.

Eco, U., 2011. *Confessions of a Young Novelist*. Cambridge MA, and London: Harvard University Press.

Ellis, B., 2001. *Aliens, Ghosts and Cults: Legends We Live*. Jackson: University Press of Mississippi.

Festinger, L., 1957. *A Theory of Cognitive Dissonance*. Stanford CA: Stanford University Press.

Freedberg, D., 1989. *The Power of Images: Studies in the History and Theory of Response*. London and Chicago: University of Chicago Press.

Gadamer, H-G., 1980. Religious and poetical speaking. In A. M. Olsen, ed., 1980. *Myth, Symbol, and Reality*. Notre Dame: University of Notre Dame Press.

Gell, A., 1998. *Art and Agency: Towards a New Anthropological Theory*. Oxford: Clarendon Press.

Gell, A., 1999. *The Art of Anthropology: Essays and Diagrams*. Edited by Eric Hirsch. Oxford and New York: Berg.

Harrison, T., 2011. Madonna Mania, *Fortean Times*, [online]. Available at: <http://www.forteantimes.com/features/articles/6154/madonna_mania.html>. [Accessed 28 September 2013].

Hartshorne, C. and Weiss, P., 1931. *Collected Papers of Charles Sanders Peirce*, Vol 1. Cambridge, Mass.: Harvard University Press.

Herbermann, C.G., 1913. *The Catholic Encyclopaedia*. 12: 85. [online]. Available at: <https://archive.org/details/catholicencyclo02commgoog> [Accessed 20 January 2014].

Hynes, E., 2008. *Knock: The Virgin's Apparition in Nineteenth-Century Ireland*. Cork: Cork University Press.

Irving, W. 1820 [2008]. *The Legend of Sleepy Hollow*. [online]. Available at: <http://www.gutenberg.org/files/41/41-h/41-h.htm> [Accessed 20 January 2014].

Irving, R., 2014. *Playing Puck: A Study of Performativity in the Shaping of a 'Legend Landscape'*. PhD University of the West of England.

Jamie, K., 2012. *Sightlines*. UK: Sort of Books.

Jung, C.G., 1959 [1987]. *Flying Saucers: A Modern Myth of Things Seen in the Sky*. London: Ark

Kleber, J.E., 2000. Pope Lick Monster. In J.E. Kleber., ed., *The Encyclopedia of Louisville*. Lexington KY: University of Kentucky Press.

Latour, B., 1993. *We Have Never Been Modern*. Boston MA: Harvard University Press.

Lennon, J and Foley, M., 2000. *Dark Tourism. The Attraction of Death and Disaster*. London: Continuum.

Locher, D., 1999. Unacknowledged roots and blatant imitation: Postmodernism and the Dada Movement. *Electronic Journal of Sociology*. [online]. Available at: <http://www.sociology.org/content/vol004.001/locher.html> [Accessed 6 April 2013].

Morgan, D., ed., 2010. *Religion and Material Culture: The Matter of Belief*. London and New York: Routledge.

Morgan, D., 2012. *The Embodied Eye: Religious Visual Culture and the Social Life of Feeling*. Berkeley CA: University of California Press.

Parker-Pearson, M. 2012. *Stonehenge: Exploring the Greatest Stonehenge Mystery*. London: Simon and Schuster.

Pearson, M. and Shanks, M., 2001. *Theatre/Archaeology*. London and New York: Routledge.

Peuckert, W-E. 1965. *Sagen*. Munich:, E. Schmidt.

Pflock, K. and Brookesmith, P., 2007. Of time and the river. In K. Pflock, ed., 2007. *Encounters at Indian Head*. San Antonio and New York: Anomalist Books.

Pryor, F., 2004. *Britain BC: Life in Britain and Ireland before the Romans*. London, Harper Perennial.

Pryor, F., 2010. *The Making of the British Landscape: How We Have Transformed the Land, from Prehistory to Today*. London: Allen Lane.

Ramshaw, G., 2014. A Canterbury tale: imaginative genealogies and existential heritage tourism at the St Lawrence ground. *Journal of Heritage Tourism*, 9, 257–69.

Reader, I., 2011. Pilgrimage, Travel and Tourism in European and Global Contexts. Talk given at *Pilgrimage in a Changing Europe: Cultural, Political and Economic Processes*, Department of Geography, University College London, June 2, 2011.

Robb, J.G., 1998. The 'ritual landscape' concept in archaeology: a heritage construction. *Landscape Research*, 23(2), 159–74.

Schechner, R., 1993. *The Future of Ritual: Writings on Culture and Performance.* London and New York: Routledge.

Schnabel, J. and Irving, R., 1992. Rolling Their Own. *Independent* magazine. 29 August. 38: 36–8.

Sharpley, R. and Stone, P.R., 2009. (eds) *The Darker Side of Travel. The Theory and Practice of Dark Tourism.* Bristol: Channel View Publications.

Sheller, M., 2011. Mobility. *socio*pedia.isa [online]. Available at: <http://www.sagepub.net/isa/resources/pdf/Mobility.pdf> [Accessed 2 August 2013].

Smith, P., 2012. *Counter-Tourism: The Handbook.* Axminster: Triarchy Press.

Stevens, W., 1945. Description without place (poem). *The Sewanee Review*, 53(4): 559–65.

Stevens, W., 1951. *The Necessary Angel: Essays on Reality and the Imagination.* New York: Vintage.

Taussig, M., 1987. *Shamanism, Colonialism, and the Wild Man: A Study in Terror and Healing.* Chicago: University of Chicago Press.

Turner, E., 2012. *Communitas: The Anthropology of Collective Joy.* New York: Palgrave Macmillan.

Turner, V., 1969. *The Ritual Process: Structure and Anti-Structure.* New Brunswick NJ and London: Aldine Transaction.

Turner, V. and Turner, E., 1978. *Image and Pilgrimage in Christian Culture.* New York: Columbia University Press.

Warner, M., 2006. *Phantasmagoria.* Oxford: Oxford University Press.

Which?, 2013. World's best heritage sites revealed by *Which? Which?* [online] 10 January 2013. Available at: <http://www.which.co.uk/news/2013/01/worlds-best-heritage-sites-revealed-by-which-304494/> [Accessed 15 March 2013].

Chapter 7

The 1652 Country

Pamela Richardson

Introduction

This chapter explores the seventeenth century origin of a pilgrimage in the north west of England, in an area known as 'The 1652 Country', and the mobility and meaning pilgrims find there today. The subject is examined against the background of recent geographical research, which explores nuances of mobility and renewal and the concepts of surface and space integral to pilgrimage (Sheller and Urry, 2004; Cresswell and Merriman, 2011; Maddrell and della Dora, 2013a, 2013b). The essence of pilgrimage is mobility; pilgrims today are widely acknowledged as people travelling on a religious quest. Geographers Tim Cresswell and Peter Merriman (2011) suggest that mobility has always been a central focus in human geography, and they quote Crowe (1938: 14): 'movement implies three essentials – origin, destination and an effective will to move'. Related to this subject are the concepts of liminality and *communitas* in pilgrimage, which is useful for setting out the sequence that occurs when people alter a hitherto stable state of belief. Uncertainty is followed by a liminal period in which change takes place and that creates *communitas*, an intense sense of community, among those who have adopted the new belief (Van Gennep, 1909 [2004]; E. and V. Turner, 1978 [1995]; Eade and Sallnow, 1991; and Coleman and Eade, 2004). Eade and Sallnow (1991) were critical of Van Gennep's concept of '*communitas*', but later scholars have found it useful. These concepts will be examined in the account of George Fox's travelling, together with a discussion of the itinerant nature of his mobility, that of his Quaker followers and other nonconformists.

A brief history of George Fox's early life explains the origin of his 'will to move', and describes the life changing effect of the two different sacred journeys he made, the first on him, and the second on the people he sought to convert. To provide physical and historical context, the chapter describes the landscape of the second journey. Elements of landscape, space and place can be thresholds into enlightenment and lead to a deeper understanding of faith and self, having the same uplifting effect as grand buildings, art and music, icons, relics and the written word. The character of the places he visited and people living there in the seventeenth century is discussed, to suggest why they were receptive to Fox's preaching. The chapter concludes with an assessment of the legacy George Fox left for pilgrims who followed his path.

George Fox (1624–1691): The Biographical Context

George Fox was born into a Christian family in Fenny Drayton, Leicestershire in 1624. As he grew up, he became increasingly disillusioned by the behaviour of people who called themselves Christians. He was also concerned about the power the Church had over people's lives, and the notion that university education was a prerequisite for training to be a minister. On leaving school, he was apprenticed to a shoemaker, who also kept sheep and cattle. His work experience was well fitted to rural life, but this was not enough for him, and in 1643, aged 19, he left home to seek a religious truth that made sense to him. It is clear from his personal journal (Penney, 1962), that geographical destinations were secondary to the people he met, debated with and questioned. For four years (1642–1646), and in spite of the civil war, Fox travelled widely, living frugally and without physical comforts. He was not the first dissenter to do this. Examining Protestant life in early modern England, N.H. Keeble points out that 'early English Protestants were forever on the road ... seeking out signs and directions in order to return to the way' (Keeble, 2002: 241). Keeble cites a contemporary of Fox, John Bunyan (1628–1688), who published *The Pilgrim's Progress* in 1685. This book, with its figurative use of landscape, destinations and people, could be understood as an allegory of George Fox's personal journey. He encountered in reality the challenges, hardship and respite that the pilgrim Christian learned from the landscape and people in the book. Mountains of doubt and mistrust, and delectable mountains offering space and vision, meadows of calm, valleys of humiliation, narrow paths of obedience, quagmires and ditches of despair and uncertainty (Bunyan, 1685 [1998]). The treacherous sands of Morecombe Bay, travelled by Fox, compare with Christian's experience of the Slough of Despond, and Bunyan, like Fox, notes that religious talk means nothing without practice (Bunyan, 1685 [1998]).

The threefold paradigm of liminality advanced by Arnold Van Gennep (1909 [2004]), explored and developed further by Victor and Edith Turner (1978 [1995]), can equally be read as a metaphor for George Fox's own slow progress. He moved from dissatisfaction with the existing pattern of religion, to enlightenment and acceptance of new ideas. Van Gennep stressed that men and women resemble nature, in which neither the individual nor society is independent. There is transition, movement and inactivity in nature, which is echoed in human lives and thresholds from one state to another, can be physical, ceremonial or territorial. Thresholds, for George Fox, were physical, territorial and spiritual. He moved through the landscape searching for answers, finding revelation in places and people, and in his own understanding. A series of 'gateways' led to his new life, a personal relationship with God, and an inner sacred place; a source of strength and a refuge from the ridicule and persecution that he and all early Quakers suffered.

By 1647 he felt ready to become a spiritual leader and set out to offer people a space of renewal in which they could find the same 'sense of spiritual fulfilment' (Maddrell and della Dora, 2013a: 2). On his second journey, as a preacher, his thresholds were still physical, such as reaching the summit of Pendle Hill, or finding the destinations he sought and the spaces where he could see his visions. He invited communities to share his experience and attain that feeling of social equality, togetherness and solidarity with people who wanted something of value to believe in. Those who followed him created those intense feelings of *communitas* that bound a growing circle of Friends. 'The heroic evangelistic journeys undertaken by Fox after his convincement are legendary.'(Keeble, 2002: 248). By no means the first itinerant preacher of the period, in his own words he 'travelled up and down as a stranger on the earth, which way the Lord inclined my heart' (Penney, 1962: 8). N.H. Keeble quotes many examples of wandering preachers and suggests that 'Restoration persecution compelled nonconformists … to follow … itinerant ministries' (Keeble, 2002: 248). Their motives also included a compulsion to spread the Word, and in this there is an echo of the mobility of the first disciples and later apostles of Jesus Christ.

Fox's interpretation of Christianity was a simple structure of belief and trust in God, with self-discipline, freedom of conscience and equality, but without paid ministry, sacraments or ritual. There was no ceremonial threshold to admit believers. His conviction was that people must escape from artificial religion of dogma and doctrines, constructed, he said, like a new kind of Babel Tower in the hope that God could be reached by climbing up a pile of theological notions and theories; a view shared by John Bunyan. Fox taught people to listen to God and build a personal relationship with Him, find confidence in their own 'inner light' and live by it faithfully and honestly, in tune with nature and with an awareness of the needs of others. It was a time of renewal for his many followers.

He emphasised the equality of the sexes before God. In one place he recounted, 'I met a sort of people that held that women have no souls … no more than a goose, but I reproved them, and told them that was not right' (Penney, 1962: 7). He moved the focus of faith from the man-made pomp of church and cathedral into the greater grandeur of God's creation and the wider landscape of a church of people. He was ridiculed, vilified, and attacked, but he never faltered in his beliefs. He continued to travel through the countryside and into many towns, including London, challenging the authority of the church to charge for ministry and impose tithes, and seeking to correct their interpretation of scriptures. When driven out of towns and villages he slept often in open fields or ditches (Penney, 1962). His first imprisonment was in 1650, for a term of six months. At the trial Judge Bennett labelled him and his followers 'Quakers' because they bade listeners tremble at the word of God. News of his preaching was spreading, and men in authority were becoming anxious and angry about its effect. At the beginning of 1652, a warrant was issued against him preaching throughout the West Riding of Yorkshire. He and his followers continued into East Lancashire and there:

> We came near a very great hill, called Pendle Hill, and I was moved of the Lord
> to go up to the top of it; which I did with difficulty, it was so very steep and high.
> When I was come to the top, I saw the sea bordering upon Lancashire, and there,

> on the top, I was moved to sound the day of the Lord and the Lord let me see in
> what places He had a great people to be gathered. (Penney, 1962: 59–60)

This vision began his ministry in the 1652 country. The American writer and
theologian Rufus M Jones (1863–1948), one of the most influential Quakers of the
twentieth century, wrote in 1924:

> Two spiritual leaders have emerged in our generation into a new light and have
> become greatly intensified living forces. They are St Francis of Assisi and
> George Fox of Fenny Drayton … Now at length they have been discovered
> as they really were in the local colour and historical setting of their own time,
> George Fox as a genuine man engaged in a dramatic struggle for reality and truth
> (cited in Penney 1962: xi).

When he died in 1691, George Fox was 67 years old and in poor health after
his relentless persecution. He was buried in a Quaker burial ground, adjacent to
Bunhill Field cemetery in London, which had been purchased by the Society of
Friends in 1661. His teaching has survived, and in the present century the tenets of
Quakerism remain largely unchanged.

Cumbria and Westmorland: The Place and its People

Landscape can have a profound emotional and mental effect on humans, which
is often at the heart of religious feeling. As Maddrell and della Dora write, 'It
captures aspects of the environment experienced directly through the senses; ways
of seeing and physical interaction with the land, movement and contemplation'
(Maddrell and della Dora, 2013b: 1106). Height is associated with proximity to the
Almighty: He is God 'above', in position and in power. It can be awe-inspiring, but
also frightening, due to the isolation and the danger of falling. Deep cool valleys
offer shelter, protection and sustenance, or give rise to feelings of constriction and
lack of freedom.

The landscape of Cumbria and Westmorland is a dramatic surface, which on
closer inspection reveals a colourful interior of places and people. The writer
Daniel Defoe (1660–1731), during a summer visit, compared the high and wild
hills of Cumbria with the South American Andes (Hoskins, 1955 [1985]). Its
hills, with their spectacular views, contrast strongly with the gentle folds of the
dales and deep, quiet valleys with rivers and streams. In the seventeenth century,
Westmorland was largely uncultivated, although in much of this part of England
the great forests were increasingly being felled, to create a field system of small
farms, joined by paths and lanes, which was changing the surface of the landscape
(Hoskins, 1955 [1985]). Mark Richards, in the twenty-first century, describes this
part of Cumbria as:

... an amazing place to walk, with ewes and lambs in almost every field there is fun and hope in the air ... a delightful valley stretch of the Dale. Then climbing above the woods on the west side of the Lune gorge, admire the sleek-lined Howgill Fells and visit the site of George Fox's impromptu pulpit. (Richards, 2004: 44)

Water is also part of landscape. Maddrell and della Dora write that 'Renewal and restoration are at the heart of Christian theology and water is central to both its offices and symbolic narratives' (Maddrell and della Dora, 2013a: 5). Rattue (1995) suggests that crystal-clear water bubbling out of the ground must have appeared magical to early man, and there are still claims of miraculous cures. In Great Britain and Ireland there is a long history of holy wells and springs as therapeutic places of pilgrimage and the work of Ronan Foley (2010) discusses the continuity between the pre Christian and contemporary Christian era. Other religions consider water vital to their belief. In Hinduism rivers have a special spatial and social dimension (Mohan, 1983), and early Vedic teaching is that the Ganges is a river goddess purifying the living and the soul after death, helping the pilgrim to attain salvation (Prideaux and Cooper, 2009).

In seventeenth-century England, however, holy wells were declining as places of pilgrimage, and rivers were not regarded as sacred, but more as providing life, health and mobility for human settlement. The river Lune, which runs through the countryside for 44 miles, from Ravensdale, Cumbria, to south of Lancaster where it joins the sea, served many purposes in the 1652 country. In addition to providing water for people and animals, it was an important source of power and transport. The small towns of the Lonsdale area take their names from the Lune, whose name probably has Celtic origins, meaning pure. It is a river of changing moods, metaphorically comparable to the variability of the life course – calm, quiet water flowing through pastures becoming at one point a dramatic, turbulent cascade through a narrow gorge. George Fox drank from it and used it for washing in his outdoor life. The symbolism of washing away sin by the purifying power of faith was preached on the banks of the Lune (Penney, 1962: 70). Here, Fox gathered more followers, fulfilling his vision of a great crowd in white clothing gathered by a river. He did not regard the river as sacred and did not baptise converts in its water, for he did not look for a single ceremony of initiation but for a continual growing in the Holy Spirit.

We use elements of landscape as a part of everyday language to describe and convey meanings associated with our life journey. We talk about having a mountain of work, going through a rough patch, and speak of 'top' as the zenith, the destination to aim for – being on top of the world, getting to the top in life. The verb to flow describes mobility of people, traffic, conversation and ideas as well as water. It is difficult to ignore the influence of landscape and nature on the human psyche. It is the surface upon which our lives are etched and can change moods and health and govern life choices.

The people of this area were tough and self-sufficient, working hard in the uncompromising terrain. They lived simple lives in the rhythm of nature, and their homes were sturdy stone dwellings. Many kept sheep, to supply the thriving cloth industry based around the river Kent, at Kendal. They also reared cattle for milk and meat, kept poultry and grew crops. There was work in forestry and its associated crafts – charcoal burning, carpentry, joinery and furniture making. There were local schools, and people were aware of politics and the importance of religion. Local landowners and merchants lived in grander properties, those in the country often fortified against the threat of invasion by the Scots. In the seventeenth century, these people already exhibited features that were identified later with Quakers – plainness in speech, dress and living.

There was political volatility throughout England after the civil war. The traditional certainties had gone, and people felt powerless and bewildered about their future. There was also discontent with the power of the established church and its ministers, and active dislike of Roman Catholicism. Since the Dissolution of the monasteries by Henry VIII in 1536, people had hoped for religious change. The King ordered clergy to revise their teaching, and to preach against the worship of images and relics, extravagance, corruption, and practices deemed to be superstitious, such as pilgrimage. Change was a slow process; however, Stephen Bigger, an educationalist and anthropologist, argues that, 'Disorder and chaos can be fruitful in that new ideas and forms can emerge from it. We should value the potential' (Bigger, 2010: 3). This was evident in the north west of England, where the Westmorland Seekers were just one of those groups who were in disorder, actively seeking a new belief. Their leaders were men who had separated from all religious bodies and many were early converts to the doctrine preached by George Fox (Penney, 1962). Margaret Fell was a Seeker for many years before she met Fox. Those who were convinced by him became part of 'the social partnership process which brings people into positive relationships which encourage change' (Bigger, 2010: 9).

This area in the seventeenth century was fertile ground for a powerful preacher like George Fox. It was also dangerous, for the men in charge of church and state were aware of the undercurrents, and fearful of anyone who might incite protest. It was as much this fear of insurrection, as offence against the church, which prompted the many imprisonments of George Fox and the detention of great numbers of his followers before and after 1652.

The Mobility of George Fox and his Message

In 1652 there was little movement of people from the area in which they lived and worked. Their scope of travel was to and from work, as far as their legs would carry them and in that difficult terrain, it was not a great distance. 'Walking was destination-oriented … a functional means of transport, and shaped by economic choices and constraints' (Lorimer, 2011: 22). Those constraints put the onus of mobility on to the itinerant preacher, and the distances covered by George Fox, mostly on foot, were immense. The three essentials for mobility defined by Cresswell and Merriman (2011) above are evident not only in his personal quest, but also in the mission journey which followed, and subsequently in the on-going ministry of his disciples in the world and in the present day travels of pilgrims who have continued to follow his path.

Pendle Hill, once famous for its connections with witchcraft, was where he had his first vision. The distance between the nearest town, Clitheroe, and Kendal, where he stayed some time, is 49 miles on the roads of today, and from Kendal to Swarthmoor, Ulverstone, where he made a base, a further 25 miles. The distance round Morecambe Bay to reach Ulverstone was long, compared with the short but treacherous low-tide journey across Morecambe Bay whose benign surface covered dangerous quicksand. Negotiating the difficult terrain, the number of miles he travelled on foot was considerably more. Prompted by his vision on Pendle Hill, Fox moved northwest through the dales, crossing countryside which seemed to echo the highs and lows of life itself.

It was at Sedbergh, famous for its grammar school founded by Roger Lupton (later to become provost of Eton), in the reign of Henry VII (1485–1509), that he found the subject of his vision the 'great people to be gathered'. This would become a key component of Quakerism:

> The experience of being 'gathered' by God leads into the experience of being guided by God. This was not just the experience of individuals, important though this is. The key to the development of Quakerism is the understanding of corporate guidance which tests and informs individual leadings. At the heart of this is the meeting for worship where Christ, the Inward Light, is present and is met. (Yearly meeting of the Society of Friends, 2004: 19.19)

Three miles north of the town, at Firbank Fell, he preached from a rock pulpit to over a thousand folk eager to hear him. As Richards notes:

> In the story of Quakerism, Firbank Fell is as near as one comes to a shrine, reflecting the personally rooted faith. The situation itself rock-exalted, exposed to the elements from every front. A place where the spirit of man and the elements may come together in a natural harmony, stripped of pomp, pride and politics. (Richards, 2010: 1)

From that gathering at Sedbergh came the group of men and women known as 'The Valiant Sixty', whose mobility carried the Quaker message across England, the continent and the world, offering spaces of renewal to countless people. The change Fox stirred up created its own mobility, 'highlighting the role of spiritual places and practices in therapeutic environments and experiences' and bringing 'an understanding of particular places, spiritual beliefs, practices and experiences, and notions of wellbeing, healing and renewal into more nuanced dialogue with each other' (Maddrell and della Dora, 2013a: 3). Meetings were broken up and moved elsewhere, the message was carried far away, and places became known for events which happened on George Fox's journey. Friends used their mobility to survive and to spread their faith. Within the area he visited, small meetings sprang up in the homes of followers, from which meeting houses would follow. There was an immediate need for burial sites, which were given generously. Quakers were refused interment in consecrated ground so a place of burial among a community of Friends, as they had lived, ensured continuity of belief. These were not 'memorial places' (Maddrell and Sidaway, 2010), however, for early Quakers did not have gravestones.

George Fox once preached at Preston Patrick chapel, where Seekers met monthly. Thomas Camm, a local twelve-year-old, was present at that meeting. He became a Friend, and was later sued for non-payment of the church tithe in that same building, which was also used as a courtroom. Possibly the history of Pennsylvania began at his home 'Camsgill', for Thomas's father John travelled in the Quaker ministry and convinced Thomas Loe of Oxford, who in turn convinced William Penn who founded Pennsylvania, USA (Foulds, 1952).

Nonconformist meetings were illegal under the Conventicle Act of 1670, but all Quaker meetings were open to everyone. If those attending were caught they served time in prison, but they went cheerfully and they and their families were supported throughout by other Friends. William of Orange took the throne from James II in 1689, and passed the Act of Toleration in the same year, giving freedom of worship to all. However, there was continuing prejudice and persecution for many more years, particularly in rural areas.

Occasionally followers accompanied Fox along the way, which provided opportunities to expound his message, answer questions and enjoy debate. Others, told he was coming, travelled to meet and welcome him. To match his physical mobility, he found energy in the thinking of those who were eager to hear him. They were prepared to change direction, mentally, spiritually and in their way of life. It took courage to be different, as George Fox knew well. Sometimes, men of authority would come to challenge him, or prevent his entry into a building or town on the pretext that he was breaking the law. When he did disturb the peace or upset a congregation, he was driven away with sticks and stones. It was an arduous progression.

There are three Yealand villages in this area: Yealand Conyers, Yealand Redmayne and Yealand Storrs. Richard Hubberthorne, a local man who had been a captain in Cromwell's army, was convinced by George Fox and became one of the Valiant Sixty. He died of malnutrition and ill-treatment in Newgate prison 10 years later. Robert Widders, another Friend from Yealand meetings accompanied George Fox when he travelled in America from 1671–1673 (Penney, 1962). The original Day School, built at Yealand in 1709, has since become a hostel for tourists and pilgrims during the summer months. Kendal was a fruitful place of ministry for George Fox. His many converts there enabled a thriving meeting to grow and the large early meeting house was rebuilt in 1816. The town is now a main hub of the region's tourist industry, so it attracts many tourists in addition to pilgrims.

Swarthmoor Hall at Ulverston was the home of Judge Thomas Fell (1598–1658). His wife Margaret Fell, née Askew (1614–1702), was generally acknowledged to be related to Anne Askew (1521–1546), the Protestant martyr, who offended society and the church by standing up for women's rights and was burned at the stake by Roman Catholic officials in 1546 (Beilin, 1997). When George Fox first visited Swarthmoor in 1652, Margaret received him (as her husband was away), and was impressed by his preaching. She persuaded him to join the congregation at Ulverstone church the following day, to enable her to compare his words with those of the priest. That convinced her of the power and sense of Fox's message, and she and her daughters became Friends, (Yearly Meeting of Friends, 2004: 19.07). Thomas Fell did not join the Friends, but he was sympathetic to their cause and allowed them to hold meetings at Swarthmoor. On a further visit to Ulverstone, later that year, George Fox became involved in an altercation with the priest and congregation and was severely beaten. One of his hands was badly injured and before them all, he prayed to God for healing and the hand was restored, 'I looked at it in the love of God and in a minute I recovered' (Penney, 1962: 73). He demonstrated healing power among his followers on many occasions and his gifts of mental and physical empathy and encouragement were certainly part of a healing process but the 'Book of Miracles' performed by him, could be interpreted today as a series of healings by naturalist medical cures.[1]

From the beginning of her convincement Margaret worked tirelessly for the Society and for a better role for women, underlining her importance to the history of female emancipation and adding credibility to her claim of relationship to Anne Askew. In 1669, 11 years after the death of her husband, Margaret married George Fox. Together they encouraged women to play an important part in early Quakerism. Many became ministers, for he held that souls were not gendered, and that the calling to preach came to men and women equally.

Neither were women afraid to travel in their ministry. Mary Fisher (1623–1698) was a Yorkshire servant girl, convinced by Fox in 1652 and sent to prison for witnessing her faith in a Selby church. She learned to write during her series of imprisonments and was with Friends who travelled to Turkey to proselytize in 1657. Elizabeth Hooton (1600–1672) was convinced in 1647, and imprisoned four times in eight years. She travelled to America with Fox in 1671 but passed away in Jamaica (Penney, 1962). Mary Dyer became a Quaker in America, returned to England in 1656, and went back two years later to challenge the sentence of death pronounced on any Quaker entering Massachusetts. She was hanged on Boston Common in 1660 (Quaker Tapestry Publications Committee, 1998). In the nineteenth century there were a number of women from wealthy families who travelled in the ministry, leaving children in the care of servants. (Maria Barclay

---

1   Fragments of this lost Book of Miracles were discovered by Henry Cadbury in 1932 and edited and published by him in 1948.

and Elizabeth Fry were both ministers, for example.) From the beginning Quaker women also published tracts on private presses and distributed them, for George Fox believed the written word was the best way to spread his teaching and to ensure that the message was consistent.

In June 1660, there were about seven hundred Quakers, including their leader, imprisoned on a variety of charges. Margaret petitioned Charles II for their release, which he granted. She also made clear to the King the corporate testimony of Friends against all wars. Perhaps her best-known work was a leaflet, 'Women's Speaking Justified', published in 1666 and still available today. At that time, she was a prisoner in Lancaster gaol, threatened with life imprisonment and the loss of her entire estate for allowing her home to be used for meetings and for refusing to take the oath (Yearly Meeting of Friends, 2004: 19.38). That sentence was rescinded but she was imprisoned for a further year in 1669 shortly after her marriage to George Fox. Together they established the administrative and pastoral centre for the Society of Friends at Swarthmoor. On her death in 1702, it passed out of the family but returned to descendants of the Fell family in 1912. British Quakers, with the generous help of American Friends were able to make it the home of the Society of Friends again in 1954. Here the basic structure of meetings, procedures and administration was formulated, which would define Quakerism for the world. Like all religions, it gradually became overlaid with rules and strict discipline, but the basic concept of living faith has never changed.

While Lancaster was not part of Fox's original journey, the prison played an important part in early Quakerism. Most of the Friends in the district spent weeks and sometimes years in its dark and filthy dungeons (Foulds, 1952). All were prepared to cross the boundary of liberty and place for spiritual freedom. Fox was tried three times at Lancaster Castle and between 1663 and 1666 the severity of treatment there and at Scarborough irreparably damaged his health. The first meeting house was built in Lancaster in 1677 and The Friends School which was first held in meeting house premises survives as The George Fox School.

Some people are moved spiritually by man-made creations – great buildings, religious or secular; gardens, music and art; icons and relics; which they find in a variety of pilgrimages of other denominations. However, most artistic, creative works are prompted by nature and geography of place, as well as by religious fervour. Landscape, places and things can all awe and inspire, but on a religious pilgrimage to cherish places or buildings too highly could be seen as idolatry, which is:

> ... a common worry in Protestant theory ... it is important to experience God
> through place in all its variety: native landscape and architecture, gardens, sports
> venues, pilgrimage, maps and town planning. Religious consciousness is cultural
> and artistic contexts re-invigorate debates concerning the existence of God ...
> landscape itself is potentially sacramental. (Brown, 2004 [2005]: 153, 154)

For some pilgrims, place alone is enough to restore and re-balance; for others, meaning only comes when associated with past events and the people who made that place or event special, providing a deeper connection to the religious history. Religious pilgrims do feel the need to pray where others have shown their faith.

The Legacy of George Fox: Contemporary Pilgrimage to the 1652 Country

This chapter has traced the detail of that first missionary journey by George Fox in order to explore the ways in which it echoes much of the subject matter of present day geographical research on mobilities. For Brown, these include an emphasis on crossing boundaries and on encounter, noting that 'The old Sankrit word for a pilgrim *thirthayatrika*, apparently literally means "one who wanders about, seeking river crossings", so the connection with crossing boundaries is an intimate one' (Brown, 2004 [2005]: 217), and that pilgrimage can be seen as encompassing 'A place of spiritual encounter when lives need a better or different spiritual dimension, one in which God draws the pilgrim into another reality where He is transcendent, or enters his/her reality as an immanent presence' (Brown, 2004 [2005]: 38). Meanwhile, Maddrell and della Dora stress the renewal concept at the heart of pilgrimage and 'the therapeutic nature of sacred space' (Maddrell and della Dora, 2013a: 2).

The 1652 Country continues to offer enlightenment to the pilgrims of the twenty-first century, more than 350 years after George Fox claimed the first converts to his revitalised, God-centred faith and way of living. The first recorded pilgrimage to the 1652 country was in 1891. Six years after the invention of the first safety bicycle, Ernest Taylor (1869–1955), cycled the 1652 country. In 1924, the tercentenary of George Fox's birth, Taylor guided a group of 400 people round Preston Patrick and Sedbergh after a five–day International Conference at Kendal. There was a pilgrimage of Yorkshire Friends in 1930 and regular trips were made by senior pupils at Quaker schools from 1936. Ernest Taylor was succeeded as a guide by Elfrida Vipont Foulds, whose guide book *The Birthplace of Quakerism* in 1952 coincided with an International Conference at Kendal, celebrating the tercentenary of the Quaker movement. A new guidebook, *The Cradle of Quakerism* was published by Quaker Books in 2011 (Kincaid, 2011). Currently there is an annual summer pilgrimage, which would-be pilgrims can join, or a Pilgrimage co-ordinator will help individuals to make other arrangements. There is also a Pilgrimage Cluster, a group of local Quakers, who provide support and guides in the area.

There are elements of the secular in all religious pilgrimage. Present day pilgrims come from all over the world using a variety of transport systems. Following the 1652 country pilgrimage, some simply want to relive the historic nature of Fox's journey, how his faith evolved, how his mission was received and carried forward. Feminists can celebrate his insistence on the equality of sexes in that early period of history, the example of Margaret Fell and the important role of women in Quakerism. Those interested in human rights issues can visit

Lancaster Gaol and learn of the ferocious religious persecution which was rife in England in the seventeenth century and which stills exists in the world today. Geographers can look at the surfaces travelled, the routes and landscapes that George Fox followed and compare them with the density of development and roads and railways nowadays. Travelling is easier, faster and more comfortable now but with enough time, many pilgrims prefer to travel on foot, for the journey is part of the experience. Being taxed physically but not giving up, experiencing in small measure what George Fox and his followers undertook, can deepen the spiritual as well as the physical meaning of pilgrimage, taking home the surface fragments of blistered feet and guidebooks as reminders of the path trodden (Maddrell and della Dora, 2013b).

The main sites for pilgrims are Pendle Hill, Sedbergh and Firbank Fell. Kendal, also a popular tourist destination, has delightful lanes and ancient yards, and an imposing Georgian Quaker meeting house where the Quaker Tapestry can be seen. Here, pilgrims can learn and marvel at the history and outreach of the Society of Friends shown in 77 embroidered panels, illustrating their faith and diversity. The panels were worked by Quakers worldwide. The project, begun in 1981, was masterminded by Anne Wynne-Wilson. More intrepid pilgrims can choose the difficult route over the treacherous sands of Morecambe Bay to reach the ultimate destination of George Fox's 1652 mission, Swarthmoor Hall, and experience personally the same welcome and hospitality that, as a travelling minister he would have received (Kincaid, 2011: 55).

Today, via the internet or film, it is possible to go on a virtual pilgrimage at any time and that has value for those who are unable to travel. However, nothing can replace the intensity of the lived experience, the sights, sounds and smells beneath the landscape surface, the atmosphere and fellowship. This pilgrimage offers no set route, no grand buildings and personal relics, no statues or Stations of the Cross, religious paintings or shrines at which to leave votive offerings. It is a simple journey through wonderful countryside to relive moments of history, stopping at farms and old stone buildings, redolent with echoes of silent worship; no hymns but an abundance of prayer, and the quietness in which the voice of God can be heard. In these places, the pilgrim creates his/her own shrines, where the necessity to pray is most deeply felt, and there is a need to reflect on the example of the faith of earlier pilgrims and marvel at their strength and courage.

Pilgrims find as great a variety of meanings in this pilgrimage as there are reasons for being there. Rewards are found in fellowship, in the value of honesty and lack of ornamentation, respect for the equality of all and remembering what peace there is in silence. But the fundamental truth that George Fox taught is that faith is a simple, uncomplicated relationship between God and man or woman, whoever the subject, whatever the religion and wherever the places and spaces one finds it. True faith is not just a surface, an overcoat to be worn on Sunday, it is the foundation of daily lives well lived, in communion with others and within the balance of nature.

If pilgrims to the 1652 country in this century find renewal in the meaning and comfort of that simple fact and mobilise it in a world that is becoming increasingly complex and challenging, then George Fox's first journey has fulfilled his visions and his message is still relevant. As Turner and Turner note, 'It is true that the pilgrim returns to his former mundane existence, but it is commonly believed that he has made a spiritual step forward' (Turner and Turner, 1978 [1995]: 15).

References

Beilin, E.V., 1997. *The Examination of Anne Askew.* Oxford: Oxford University Press.

Bigger, S., 2010. *Thresholds, Liminality and Fruitful Chaos: Revolutionary Change in Education?* [e-book] British Education Studies Association. Available at <http://eprintsworcs.ac.uk/834/1> [accessed 27 January 2011].

Brown, D., 2004 [2005]. *God and Enchantment of Place: Reclaiming Human Experience,* [e-book] Oxford: Oxford University Press. Available at <http://www.oxfordscholarship.com/view/10.1093/0199271984.001.0001/acprof-9780199271986> [Accessed 15 May 2013].

Bunyan, J., 1685. [1998] *The Pilgrim's Progress.* Oxford: Oxford University Press.

Coleman, S. and Eade, J., 2004. *Reframing Pilgrimage: Cultures in Motion.* European Association of Social Anthropologists, London: Routledge.

Cresswell, T. and Merriman, P., 2011. *Geographies of Mobilities: Practices, Spaces, Subjects.* Farnham: Ashgate.

Crowe, P.R., 1938. 'On Progress in Geography 'in Scottish Geographical Magazine, 54(1), 1–18.

Defoe, D., 1928. *A Tour through England and Wales.* London: J. M. Dent and Sons Ltd.

Eade, J. and Sallnow, M.J., 1991. *Contesting the Sacred: the Anthropology of Christian Pilgrimage.* London: Routledge.

Foley, R., 2010. *Healing Waters: Therapeutic Landscapes in Historic and Contemporary Ireland.* Farnham: Ashgate.

Foulds, E.V., 1952. *The Birthplace of Quakerism: A Handbook for the 1652 Country.* London: Friends Home Services Committee.

Hoskins, W.G., 1955. [1985] *The Making of the English Landscape.* London: Penguin Books.

Keeble, N.H., 2002. 'To be a pilgrim': constructing the Protestant life in early modern England, in C. Morris and P. Roberts, eds, *Pilgrimage. The English Experience from Becket to Bunyan.* Cambridge: Cambridge University Press, 238–56.

Kendal and Sedbergh Quakers, n.d. *History of Quakerism in this Area,* available at: <http://www.kendal-and-sedbergh-quakers.org.uk/page12 html> [Accessed 15 June 2013].

Kincaid, A., 2011. *The Cradle of Quakerism .Exploring Quaker roots in North West England.* London: Yearly Meeting of the Religious Society of Friends.

Lorimer, H., 2011. Walking: New forms and spaces for studies of pedestrianism, in T. Cressman and P. Merriman, eds, *Geographies of Mobilities – Practices, Spaces, Subjects.* Farnham: Ashgate, 19–33.

Maddrell, A. and della Dora V., 2013a. Editorial: Spaces of renewal. *Culture and Religion: An Interdisciplinary Journal,* 14(1), 1–7.

Maddrell. A.and della Dora, V., 2013b. Crossing surfaces in search of the Holy. Landscape and liminality in contemporary Christian pilgrimage. *Environment and Planning A,* 45(5), 1105–26.

Maddrell, A. and Sidaway, J.D., 2010. *Deathscapes. Spaces for Death, Dying, Mourning and Remembrance.* Farnham: Ashgate.

Mohan, B.S., 1983. *Hindu Places of Pilgrimage – India: A Study in Cultural Geography.* California USA: University of California Press.

Penney, N., 1962. *The Journal of George Fox.* London: J.M. Dent and Sons Ltd.

Prideaux, B. and Cooper, M., 2009. *River Tourism in the South Asian Subcontinent.* Wallingford: CAB International.

Quaker Faith and Practice: The Book of Christian Discipline of the Yearly Meeting of the Religious Society of Friends (Quakers) in Britain, 2004. London: The Yearly Meeting of the Religious Society of Friends (Quakers) in Britain.

Quaker Tapestry Publications Committee, 1998. *Pictorial Guide to the Quaker Tapestry.* Kendal: The Quaker Tapestry at Kendal.

Rattue, J., 1995. *The Living Stream: Holy Wells in Historical Context.* Woodbridge: The Boydell Press.

Richards, M., 2010. *44 Firbank Fell* [online], available at <http://www.markrichards.info/walk-free/park-and-stride_1/park-and-stride-41–50/44-firbank-fell.html> [Accessed 15 January 2014].

Sheller, M. and Urry, J., 2004. *Tourism Mobilities, Places to Play, Places in Play.* London: Routledge.

Turner, V. and Turner, E., 1978, [1995]. *Image and Pilgrimage in Christian Culture.* Columbia, NY: Columbia University Press Paperback.

Van Gennep, A., 1909, [2004]. *The Rites of Passage.* London: Routledge.

Chapter 8

The Way of Saint James: Memory, Propaganda and Power

Belén Mª Castro Fernández

Introduction

The Way of Saint James is based on a secular tradition that has transcended its religious origins, becoming a highly attractive tourist product and an important hallmark of Galician identity in North West Spain. Its significance cannot be understood without reference to the figure of St James the Great, who, according to tradition, preached throughout the Iberian Peninsula, and, in the year 42 AD, was beheaded in the lands of Jerusalem under an order from Herod Agrippa. His disciples carried his body and head by boat from Jerusalem to Iria Flavia where, with the help of Queen Lupa, he was finally laid to rest.

Due to the secrecy of early Christian society, the location of the tomb remained a mystery until the ninth century, during the reign of Alfonso II of Asturias and Galicia. At some point during this period, a hermit named Pelagius saw lights in a place called Libredón and reported this to Bishop Theodomirus. As a result excavations were organised and this led to the discovery of the tomb containing bodily remains which were identified as the Apostle. Consequently, Alfonso II ordered the construction of the first church of Santiago, and this saw the start of the pilgrimages to the west of Spain (López Alsina, 1993). This influx of people from various parts of Europe eventually formed what is known today as the Camino de Santiago or the Way of Saint James.

The final impetus for the pilgrimage to Compostela was the granting of indulgences by Pope Alexander III in 1179, consisting of the absolution of sins for those who fulfil the penance of walking the Way, and spiritual salvation when performed in a Holy or Jubilee Year, (which occurs every time the feast of St James, the 25th July, falls on a Sunday). This started a golden age (López Alsina, 2013:199–201) in which the Way's popularity helped to improve communication and led to an economic boom in towns along the *Camino*'s routes (González and Medina, 2003; Murray and Graham, 1997). The Jacobean pilgrimage has since become a cultural product that has undergone a transformation into its present function as a key tourist resource throughout northern Spain.

During the Middle Ages, the population was predominantly rural. The Way was of crucial importance as an impetus for the development of new towns such as Ligonde, Hornillos Del Camino, Rabanal Del Camino, Canfranc and Burguete. It

**Figure 8.1   Cathedral of Santiago de Compostela in the middle of the
twentieth century – main façade, Portico of Glory and
archaeological exploration**

Source: Archive Pons-Sorolla, Madrid, Spain, c. 1945–1960.

also contributed to the development of many villages such as Burgos, Redecilla Del
Camino, Cacabelos, Molinaseca, Portomarín, Puente la Reina, Santo Domingo de la
Calzada, Estella, Sangüesa or Villafranca del Bierzo. Economic needs determined
the establishment of merchant groups and craftsmen who, together with attending
to basic human needs, assisted walkers, carried out monastic orders and used royal
powers to build bridges, hospitals and welfare provisions along the Way.

However, in the seventeenth century the number of pilgrimages began to steadily
decline from its medieval peak. The decline possibly began with the disappearance
of the relics of the apostle in 1589, when they were hidden to prevent looting by
pirates. The religious wars of the early modern period, prompted by a succession
of reformations and counter-reformations across Europe and resulting in a deeply
divided Christianity, also contributed to the decline in pilgrimages, as did a series
of plagues throughout this period. Alongside the opening of alternative new trade
routes during this period, the steady decrease in crowds flocking to Compostela
resulted in a gradual degradation of the infrastructure created for the Way, and
the loss of its strategic and commercial importance. By the nineteenth century,
the flow of pilgrims had ebbed to a trickle. The industrial revolution and ongoing
social liberalisation had caused, to varying degrees, the breakdown of traditional
values. Nineteenth century society, in the process of secularisation, did not seek
solace for its spiritual problems; valued monuments were no longer churches or
religious statues, but factories or railway stations. The recovery of the Way in the
later nineteenth century, a process known as the Second Inventio, coincided with
the wider revival of pilgrimage across Europe, for which the advent of pilgrimage
to Lourdes (from 1873) had provided a catalyst.

Interest in the Way re-emerged after the appearance of the relics of San Francisco in Assisi in 1818. This was linked to a growing desire for a more rigorous historical knowledge of the evolution of the Church, with the Vatican Archives opened to researchers in 1883. This intellectual climate, together with the influence of positivist doctrines, led to the archaeological rediscovery of the relics of St James in 1879 (Otero Suarez, 1999). In 1884 Pope Leo XII issued a papal bull to authenticate the relics for use as a mechanism for Jubilee indulgences that had originally been established in the Middle Ages. With this new support, the Jacobean pilgrimage to Compostela was reborn (Villares Paz, 2004).

Spiritual Value and the Military Handling of the Way of Saint James as Elements in its Conversion into a Tourist Product

The Way's revival reached its peak in the second half of the twentieth century, when besides a religious motive, there were other interests and new social customs: these included the spirit of adventure, tourism, stress therapy and so on, which led to a significant stream of people travelling to Compostela (Lois González, 2013). Thereafter, the Way of Saint James started to show some signs of economic profitability for businesses at all points along the route, and for the church (Bermejo Lopez, 2002; González and Medina, 2003). With the arrival of the Franco dictatorship (1939–1975), the Way became an instrument of symbolic construction for the new regime which sought to manipulate and ideologically exploit the pilgrimage route and in particular the Holy Years of 1943, 1948, 1954, 1965 and 1971 (Precedo Lafuente, 1999; Rodriguez, 2004). During this period, the Church and the State became mutually supportive.

Between the 1940s and 1960s, and coinciding with the dictatorship's expansion of post-war control, the Church lent legitimacy to the regime by equating it to a crusade, one whose success was due to the intercession of St James (Compostela, 1954a). It was claimed by the Church that Spain was emerging, thanks to Franco, from an amoral stage that had begun in the Cortes of Cadiz (1810), and was entering another stage where spiritual revival and national reconstruction were to proceed hand in hand. Moreover, the Church claimed that throughout history the greatness of Spain had always been linked to the devotion of St James, to the extent that the greater or lesser popularity of the Way brought with it the splendour or decay of Spain. Franco, in turn, used this religious support to project himself as a saviour that would restore lost order in the country. From the 1960s, the dictatorship began to emphasise and exploit the economic and tourist elements of the Way of Saint James, thus moving away from the Church's ecumenical and reconciliatory message. This move enabled Franco to carry out a policy of openness on an international level. Therefore, during the Franco years there was interest in again intensifying the cult of St James in order to lift the country both materially and spiritually.

The figure of Saint James the Apostle was ideologically appropriate for the regime's chosen interpretations. Significantly, much of the recovered medieval literature presented him as a knight and warrior, interceding in favour of the Christian militias, and specifically as Saint James the Moor Slayer (Mouriño Lopez, 2001; Márquez Villanueva, 2004; Domínguez García, 2009). The liturgical ceremonies at the opening and closing of each Holy Year were enveloped in speeches and paraphernalia for the benefit of the Franco dictatorship. Indeed, the Church not only equated Franco with historical dynasties which, in its opinion, contributed to the greatness of Spain – Ramiros, Alfonsos, Fernandos – but also in its desire to highlight Franco with a supernatural significance, he was identified as chosen by God to fulfil a double mission. Firstly, to safeguard the country's Catholic unity, with support from Saint James the Apostle and secondly, to establish an equitable dialogue between his authority and ecclesiastical power, as shown in the speeches made by the cardinals of Santiago de Compostela and Tarragona and also in Franco's address during the official events of the Holy Year 1954 (Compostela, 1954a; Compostela, 1954b). For his part, Franco, besides justifying his military victory through the intercession of St James and the Virgin of Pilar and Guadalupe (cults that were enhanced during his leadership), allowed the church major concessions in national education and supported the massive restoration of ecclesiastical heritage, much of it in the area of the Way of Saint James.

From the 1960s, a change in the orientation of Holy Year advertising was observed, reflecting a change in the wider Jacobean phenomenon. If, until then, there had been a focus on values of faith and nation, this now gave way to a focus on encouraging tourist interest. Indeed, the offer made to the State by the Friends of the Camino de Santiago Group, immediately after being formed in 1962, served as a reference. This stated that its purpose was to enthusiastically work on the measures taken by the government to promote or propel the updating of the Way of Saint James in order to make it a tourist route that so interested Spain (Compostela, 1962).

The first issue of the journal published by the association was entitled *Ruta Jacobea*. It dealt with the need to differentiate between pilgrim and tourist, adapting the Way to cater for the latter by enabling them to experience a more comfortable and attractive journey, without having to keep strictly to the original route. The association thought that tourists, with only 15 days of holiday, would be unable or unwilling to experience the Way in the same manner as religious pilgrims. It assumed that in the 1960s the modern pilgrim would be more concerned with speed and convenience. So, one of the urgent tasks proposed was to draw a convenient separate but conjoined 'Way' for tourists that would simultaneously cater for their needs without losing entirely the original religious feel and purpose of the route. The post-war autarky phase also witnessed the simultaneous opening up of Spain to overseas visitors, with the marketing of the Way of Saint James to a more tourist-oriented pilgrimage group. The results of the intense promotional work carried out on the eve of the 1954 Holy Year meant that the expected number of pilgrims and visitors was exceeded, with visitor numbers calculated at 700,000 (Santos Solla,

2006: 137). During this early period an Organising Board had been established, which had begun preparations to respond to three basic questions posed some two years beforehand. Firstly, how to advertise the announcement of the Holy Year in the press, radio, mailings, posters, movies and brochures? Secondly, how to prepare Lavacolla Airport at Santiago de Compostela for a greater flow of traffic? And thirdly, the issue of accommodation; in particular, how was the conversion into a luxury hotel of the Hospital founded by the Catholic Monarchs in Santiago de Compostela to be achieved? (Lois Gonzalez and Castro Fernández, 2006).

On the eve of the next Holy Year, in 1965, in which the number of visitors to Compostela was around 2,500,000, the process of giving the Way Heritage status started with the renovation of the historic city centre in Compostela, and an improvement in the care services for pilgrims during their stay in the city. To fulfil the latter task the National Patron of Santiago de Compostela was created in 1964. This was a privilege for the City of the Apostle, and a clear sign that the State understood the enormous significance of Compostela in safeguarding the city as a universal pilgrim destination (Compostela, 1964; Corriente Córdoba, 1993).

The general interest of the State towards the Way changed substantially during the 1960s. The efforts to ratify and explain its ideological and religious position gave way to the promotion of tourism. The Director General of Business and Tourism, Leon Herrera Esteban, characterised the Way as a sentimental business with universal and European significance, one in need of public intervention and the support of individuals in order to give it new life (Compostela, 1965). Although some of the 1954 advertising campaign had focused on overseas markets, this was further intensified in 1964. Likewise, various strategies were put into place, from the reprinting of the book *St James. Patron Saint of Spain* by Marquis de Lozoya, and the approval of discounts for pilgrims travelling with RENFE (Spain's National Rail Network). There were even commemorative stamps and Christmas greeting cards issued from the Ministry of Information and Tourism, with the section of the Portico of Gloria in Compostela Cathedral on the cover.

The Way became a key resource of global importance for the government. Besides the production of posters, brochures and the offer of a special prize for short films about the Way, a booklet was published on the publications of the Ministry of Information and Tourism regarding the Jacobean route, in Spanish, French, German, English, Italian, Portuguese and Dutch, and including *Noticiario Turístico: Camino de Santiago* (Compostela, 1965). It was 50 pages long and contained data collected from all the populations along the Way. The forecast of an influx of tourist-pilgrims created demand for credit to construct hotel accommodation and the provision of loans for individual homeowners who were happy to provide rooms for pilgrims. Improvements to communications were subsidised, not only to ease arrival in Compostela, but also to offer the image of a renewed Spain in its recovery process.

The conversion of the Way into a tourist product during Franco's time reached its climax with the preparations for the 1971 Holy Year, in which the official creation of the Way of Saint James was produced by the Order of March

9th. This enabled increased investment on road improvements on the Way – the widening of roadways, bridge repairs – the installation of information offices and the restoration of monuments in order to give some support to perennial tourism and to encourage piety amongst everyone who had completed the pilgrimage to Santiago de Compostela. The government's over-riding interest was, therefore, to develop the Way as a permanent tourist route and not merely as an event whose importance was confined to a Holy Year.

The tourism aspect of the Way of Saint James is controversial and there are many authors who have studied its singularities. For example, Murray and Graham (1997) highlight the potential tensions and conflicts which might arise from the commodification of heritage artefacts, as well as the necessity of sensitive management of both the integrity of the *Camino* and the meeting of the needs of pious and secular travellers. Lopez Palomeque (2000) draws attention to the *Camino's* itinerant state – established in stages and with a final destination; its linearity as a route, and its configuration from contrasted territories that, away from the collective meaning of the Way, have a singular function and generate resources as tourist attractions. He also deals with the meaning of heritage in the cities on the Way in terms of modern processes of tourist consumption, with particular attention to the role played in this area by public authorities, an economic-related theme further addressed by González and Medina (2003).

Ecclesiastical concern about the progressive secularisation of the Jacobean pilgrimage was expressed during a meeting during the 80th anniversary of the apparitions in Fatima between those responsible for the major European centres of pilgrimage: Fatima (Portugal), Lourdes (France), Loreto (Italy), Czestochowa (Poland) and Altting (Germany). Those present were concerned about preserving the original meaning of the pilgrimage as a religious one, in an attempt to break or reduce the economic interests surrounding the pilgrimage (Esteve Secall, 2005). Singul (1999) notes that the disappearance of the traditional on-foot pilgrimage from the mid-twentieth century, which had been denounced by Vazquez de Parga, Lacarra and Uria in 1948 and Justo Perez de Urbel in 1977, was exacerbated by the number of negative events that had occurred in Europe during this period: two World Wars, political fragmentation and the restructuring of Central European states after the Treaty of Versailles, the rise of fascism in Italy and Germany and the Spanish Civil War. The revitalisation of the Way during the late Franco years (1960–1975) was promoted from the perspective of *development* using modern means of transport – rail, car and plane; hence, for the most part, participation was restricted to the festivity of St James to the detriment of the walking pilgrimage as devotional expression. It was only in the 1980s that this aspect of the pilgrimage dramatically resurfaced in response to tourist motivations.

Nevertheless, despite all the promotional campaigns, the Franco Government did not promote the Holy Year of 1971 with the same intensity as those of 1954 and 1965. The end of the Franco regime in 1975 led to a marginalisation of the Jacobean theme until, from the 1980s, it became subject to a process of heritage and tourism activation for the Holy Year celebrations of 1993 and 1999. The 1970s and

1980s were marked by the oil crises, the transition from dictatorship to democracy and the new political configuration of the State. Still, from the religious sphere, the Way witnessed a singular event: the visit of Pope John Paul II, the first pope to visit the apostolic tomb. This was repeated in 1989 in front of an estimated youth rally of around 500,000 participants (Rodriguez, 2004).

The massive displacement and intense promotional campaign of the Holy Year in 1993 turned the Way into a prominent international tourist route. However, Santos Solla (1993) notes that there was a certain scepticism surrounding its consolidation into a tourism resource, despite the extra-official promotion. This was due to the fact that it still required for many, significant physical effort and free time to undertake the journey and seemed to provide relatively little economic benefit except on specific Jubilee dates or at particular locations along the route or at the 'finishing line' in Compostela.

The Monumental Definition of the Way of Saint James: Conversion into a Heritage Asset

On the eve of the Holy Year of 1965 the attention given by the government and local authorities to the Way of Saint James came to the fore. Promotional campaigns no longer focussed so much on the cult of the Apostle or the city of Compostela as a spiritual destination for pilgrims, but on the Way itself. This began to take on its own identity and was advertised as a fundamental part of the Jacobean phenomenon. The Way became valued as a spiritual journey that, in turn, gave touristic and cultural significance to Spain (Rodriguez, 2004). Indeed, in 1962 the Government had declared the French Way, from Roncesvalles to Santiago de Compostela, as a *Historic Artistic Site,* expanding the action plan that, until then, was centred on towns and monuments located within its environment.

The diversity of artistic resources and the uniqueness of regional identities that underpin the French Way, under Franco's guidance, make it a showcase of cultural values for the north of Spain, attributing meaning to the Way's identity. Its progressive heritage status has come through architectural and promotional campaigns based on this significant endowment. As Eva Mouriño Lopez (2001) points out, the Way's process of heritage activation through intervention strategies had been mandated by the public administrations since the 1980s and was not limited to the recognition of meanings of identity, it also involved the encoding and display of objects in its physical boundaries and exhaustive restoration programmes, disclosure and tourist activation.

Work on restoring the Way fell under two headings. The first included improvements for walkability and the provision of accommodation for pilgrims; the second, the restoration of monuments and refurbishment of urban landscapes that retained evocative traces of the Way such as hostels, footpaths, Jacobean iconography and so on. The confluence of both concerns brought about a plan known as the Restoration Plan for Jacobean Places, coordinated primarily by the

Directorates General of Fine Arts and Architecture and designed for overall supra-cultural assessment (Castro Fernández, 2010).

Among planned improvements we must emphasise, in its role as a receiving centre, the Urban Renewal Plan for Santiago de Compostela (1945–1965), coordinated by the aforementioned Directorates General in collaboration with the City Council (Castro Fernández, 2013). The conservation of the old town declared an Historic Artistic Site by Decree of the Ministry of National Education on 9 March 1940 was decisive in the renovation of the socio-cultural-religious Jacobean project. Besides advances in the general city services, there was renovation and updating of historical heritage. This included the rehabilitation of unique buildings like the Gothic House, which was adapted into a Pilgrimage Museum (1951–1967), and actions linked to the pilgrims' access to the Cathedral (1950–1964) – for example, repairing new tourist itineraries, signposting, farmhouse restoration, refurbishment of squares, urbanisation of the *Porta do Camino* (1953–1965), work in the *Monte del Gozo* (1964) and so on.

The need to protect monuments and the environment – the latter not only as an artistic complement but also for its defining character – promoted the aesthetic improvement of stopping points located approximately every 50 kilometres on the Way. This led to their recovery and incorporation as Historic Artistic Sites in the declaration of the Way and in turn, into a wider European cultural heritage (Chamoso Lamas, 1964).

The places chosen for the implementation of this strategy were Berdún and Sos del Rey Católico in Huesca; Sangüesa, Puente la Reina, Estella, Ujue and Viana in Navarra; La Guardia, in Álava; Najera, Briones, San Vicente de la Sonsierra, Haro and Santo Domingo de la Calzada, in Logroño; Lera in Burgos; Villarcazar de Sirga and Carrión de los Condes in Palencia; Sahagún, Castrillo de los Polvazares, el Valle del Silencio with Peñalba and San Pedro de Montes and Villafranca del Bierzo in León; O Cebreiro in Lugo, and the Monte del Gozo in A Coruña. To this list more enclaves would be added, thanks to urban renewal work developed within them. They were declared as protected and promoted to tourists as Jacobean nucleii along the Way. Serving as reference to this is the paradigmatic case of the historic centre of Tui, in Pontevedra, which was declared a Historic, Artistic, Landscape and Picturesque Site in 1967, after extensive restoration of its cathedral and beautification of its medieval area (Castro Fernández, 2003).

The restoration and monumental definition of the Way started in Franco's times and was consolidated in the last third of the twentieth century through actions which promoted its consideration as a heritage asset, resulting in more accurate ratings as a Joint Cultural Asset and an Asset of Environmental Interest (Bemejo Lopez, 2001). It is true that the evolution of the Way as a heritage asset was not limited to the recognition of identity and religious meaning. It also involved the encoding and display of the Way through its physical markers and submission to exhaustive restoration programmes, promotion, and tourist activation (Mouriño Lopez, 2001).

The transfer of tourism policy to the Autonomous Communities in the early 1980s allowed for the development of their own strategies (Herrero, 2008). The Camino became the star product in Galicia and in other communities such as Castilla-León and La Rioja it also began to occupy a more prominent place. Such actions gave way to a stage that extends to the present, in which the Way of Saint James is considered to be a Cultural Interest Asset, worthy of the title of Major European Cultural Route by the Council of Europe in 1987, a World Heritage Site by UNESCO in 1993 for its Spanish section, and in 1996 the French section, and the winner of the Prince Asturias Concord Award in 2004.

Pilgrimage and Jacobean Mobility

The European Council Declaration of the Way of Saint James as the main European Cultural Route states that it is a set of pilgrimage routes leading to the tomb of the Apostle. This declaration refers to the pathways from different European regions, which have favoured its conversion into a major cultural route within the configuration of Europe. This cultural identity was made and enabled by the existence of a European space full of collective memory and crossed by Ways capable of overcoming distances, borders and languages. During the enhancement of the Way in the 1960s, a period of Franco's greatest ideological domination and exploitation of tourism for economic renewal, the Way was linked to Spanish national identity. However, from the 1980s its European identity was emphasised (Mouriño Lopez, 2001). After this period a number of initiatives to revitalise the Way happened at international, national and regional level in order to capitalise politically on its social impact (Herreo *et al.*, 2008).

Nowadays not only is this product based on a religious phenomenon, it is also linked to a cultural journey (Lois González and López, 2012). Regardless of religious beliefs, the truth is that there is now a route that has been travelled for centuries by a great number of travellers on their way to Compostela. The pilgrimage is not an isolated incident but part of a larger system that is society. Therefore, as society undergoes change, the pilgrimage fulfils various roles (Leira Lopez, 2004), offering continuity, but also being in dialogue with and responding to change.

Firstly, it satisfies human needs as the individual is released from their daily lives and enters the spiritual world. The pilgrimage acts as an escape route and brings benefits for both physical and mental health. Secondly, it encourages communication and social cohesion in an era where individualism reigns and where face to face relationships between individuals barely occur. Thirdly, it plays an important cultural role. The Way of Saint James has been a means of sharing both ideas and artistic and wider cultural productions for centuries, which is the reason why it has been declared the main European Cultural Route. Fourthly, the pilgrimage involves a flow of people that generates the use of services and an embodied economic revitalisation, for example, in the improvement of amenities and infrastructures in the municipalities it crosses.

It is clear that the success of the Way of Saint James is caused by the frenetic pace of present society which results in the need for some time to reflect and meditate. The Way has been valued afresh as a new tourist product with multiple resources: religious, social, cultural, gastronomic, environmental and so on. Today, extra-religious motives are becoming more and more important and are, ultimately, a different form of tourism from the others that have preceded it.

The 1993 Holy Year marked a turning point in the Way of Saint James' marketing with a strategy that was preferentially directed towards national tourism. Since then, it has taken a strategic position over other products since it offers an experiential quality in which the personal experience of each individual is the key to the journey. Nevertheless, the Way has been subjected to intense tourist pressure on some parts of the route since 1993, mainly due to the ever increasing number of tourists and pilgrims. The pressure is concentrated in both time and space. The summer months and Jubilee years see a major tourist influx affecting some routes, especially the French Way, where accommodation (above all pilgrim hotels), the supply of religious services (such as the celebration of mass), cultural and monumental heritage (museums, exhibitions and/or churches) and other forms of saturation of tourism resources may be found.

The main characteristic that defines Jacobean mobility is the heterogeneity of the Way since it passes through various cities, towns and villages, uninhabited areas or natural environments with varying protection laws. In addition, the Way is under different management by private companies and various public entities. It is also a product requiring different kinds of effort depending on the means of transport chosen: on foot, wheelchair, horse, bicycle, car or bus. The demographic profile of travellers is heterogeneous with a wide age range enjoying the same product. This enjoyment varies depending upon the company chosen for the journey; there are those who choose to travel alone, those who travel with family, their partner or a group of friends. There are also various reasons for following the Way with the main ones being religious, spiritual, cultural, sport and leisure (see Frey, 1998).

The manner in which the journey is organised is not homogeneous. It has been observed that the trips can be planned by domestic or foreign travel agencies, associations, or simply individually organised by making reservations in advance by phone or online. There is also a significant group that enjoy the adventure of having nothing planned at all. The access points to the Way are not controllable since there are various routes that start in different municipalities and as such, it is difficult to regulate visitor movement. We should also add that this route can be enjoyed in sections. Nevertheless, they all have one thing in common and that is that the Way's end is Santiago de Compostela. In addition, exogenous factors come into play such as weather which affects demand due to the activity's outdoor nature.

The revitalisation of the pilgrimage phenomenon in recent decades is reflected in the upward but not constant trend in visitation since the early 1990s with high peaks matching those seen in the Jubilee years. This growth is seen in foreign and Spanish visitors, although the evolution of the first group has been more uniform and is not affected by the celebration of the Holy Years. This shows that Holy Years are not a major factor in attracting foreign tourists, unlike domestic ones. The limited knowledge of the Jacobean itself and the meaning of the Jubilee Year can go towards explaining the national and international tourists' unequal enjoyment of the Way. Moreover, this situation may be explained by the fact that the Way's marketing strategy had not been fully developed until the beginning of the twenty-first century.

In this sense, an innovative campaign was launched by the central government in 2002 under the title of 'Spain Marks'. As part of this project, there was a special series developed for Galicia: Galicia Marks ('Galicia entices' or 'be enticed by Galicia'), which was a swing away from the conventional codes in the tourism sector. It was an international campaign in which advertising in magazines and general publications were used instead of posters (Juanatey Heredia and Santos Solla, 2005).

Ever since, international campaigns for the Way of Saint James have increased, with France, Germany, Italy and Portugal being consolidated as markets where the Way has widespread acceptance. Its image of being related to religion and spirituality has led to a diversification of products associated with the Way. In turn, countries like Brazil, Japan and Poland are becoming emerging markets in which the Way is linked to religious motivations and is identified with other routes such as Kumano Kodo (Japan).

Conclusions

Undoubtedly, the Way of Saint James is a tourist product in itself. It is the result of enhancement, promotion and marketing of tourism resources that exist along the Way leading to Santiago de Compostela. A study published in 2007 by the Institute of Tourism Studies in Galicia indicated that 86 per cent of respondents considered it an attractive and interesting tourism product compared with 7.56 per cent who considered it a pilgrimage and spiritual route and not a tourism destination.

Consequently, we are no longer seeing a product that is based solely on a religious phenomenon, but one that is linked to a wider cultural experience. The result is that the Jacobean pilgrimage has now become a matter of business. As a visitor the pilgrim wants to see historic and artistic sites, to be at one with nature, stay in decent accommodation and purchase souvenirs to take back home. Hence in most cases, there is an overlap between the pilgrim and the figure of the tourist and this affects both marketing and promotion of the product.

Nowadays, to talk about the Way of Saint James is to refer to a religious behaviour, a way to seek oneself, a cultural resource, a tourist product in which the initial social-political, ideological and economic connotations have been changing over the centuries and its mainly European aspect has generally strengthened. We live in an era of globalisation, which beyond the implied international economy network shows a closer and more accessible facet addressing concepts such as multiculturalism, miscegenation, discrimination and identity. In this sense, as a tourism product, the Way promotes exchange, works against discrimination and helps us understand one another better, without any loss of identity for one or the other.

It must be added that the Way is now a popular attraction, a phenomenon on the rise which is linked to a type of mass tourism that is based on the use of route itineraries programmed by travel agencies and tour operators. These are intensively marketed, above all as in the Declaration of Heritage and as in the case of the Way and its final Jacobean destination. Although a tourist may not have walked the Way, he/she will know that Santiago is mainly a pilgrimage city and as such, will understand that before visiting.

Together with this type of tourist, there is another type with extensive higher education, who is more knowledgeable and who demands a higher quality of tourism product. This explains the rise in demand for cultural and heritage products and indicates that the Way satisfies this group's demands for knowledge, whether geographical, sociological, historic, gastronomic, sports or even personal. Amidst these new travellers, where does the original pilgrim stand? This is the question that has generated one of the most obvious conceptual confusions surrounding the Way as a product, both in how to design the product (for example, group pilgrimages organised by religious associations compared to package tours organised by travel agencies) and the manner in which it is enjoyed (whilst the pilgrims stay in hostels, the tourists choose other type of establishments such as hotels and paradores).

The figure of the pilgrim has actually undergone a change which goes beyond a modernisation of clothing. The pilgrim now has more versatile demands stemming from greater knowledge; that is to say, the pilgrim once set off with little understanding of the surroundings, the environment or the dangers to be faced and so on. Nowadays, everything is under control and the pilgrim is now developing a special role that equates to something closer to a conventional tourist. If anything separates them, it is the motivation for the journey – strictly religious (self-improvement, charity, generosity and so on) against other groups with a greater diversity of motivating factors.

Undoubtedly, pilgrims represent an interesting market in which there is no competition to contend with. From a tourist point of view, the concern mainly lies in the pilgrim's limited spending and their short stay in the city, given that most sleep in hostels or somewhere similar and after weeks of walking, tend to spend very little time in the destination city. Nevertheless, pilgrims are simultaneously tourists and therefore potential sources of revenue for the tourism industry. That is, they form part of the image of the Way of Saint James and contribute to and enrich the enjoyment experienced by conventional tourists. In this sense, the sight of tired pilgrims in the centre of Santiago de Compostela and on the Way is common in the promotional language of the Jacobean product.

References

Bermejo Lopez, M.B., 2001. *El Camino de Santiago como Bien de Interés Cultural. Análisis en torno al estatuto jurídico de un itinerario cultural*. Santiago de Compostela: Xunta de Galicia.

Bermejo Lopez, M.B., 2002. La protección de los itinerarios culturales: en especial la protección del Camino de Santiago a través de las técnicas protectoras del patrimonio histórico. *La diversificación y la desestacionalización del sector turístico. IV Congreso de Turismo, Universidad y Empresa*. Valencia: Tirant Lo Blanch, 337–51.

Castro Fernández, B.M., 2003. *La restauración de la catedral de Tui. Historicismo y conservación en la Catedral de Tui (siglos XIX y XX)*. Sada, A Coruña: Ediciós do Castro.

Castro Fernández, B.M., 2010. *El redescubrimiento del Camino de Santiago por Francisco Pons-Sorolla*. Santiago de Compostela: Xunta de Galicia, S.A. de Xestión do Plan Xacobeo.

Castro Fernández, B.M., 2013. *Francisco Pons Sorolla. Arquitectura y restauración en Compostela (1945–1985)*. Santiago de Compostela: Universidad de Santiago, Consorcio de Santiago.

Chamoso Lamas, M., 1964. Sobre los conjuntos urbanos de interés histórico-artístico existentes en el Camino de Santiago. *Boletín de la Archicofradía del Glorioso Apóstol Santiago*, 57: 15–17.

Compostela, 1954a. Alocuciones por la Emisora de Radio Nacional de Madrid, de los Emmos. y Revmos. Cardenales de Santiago y Tarragona, Dres. Quiroga Palacios y Arriba y Castro, en vísperas de la Apertura del Año Santo Compostelano. *Compostela*, 28: 2–3.

Compostela, 1954b. Ofrenda Nacional. Invocación de S. E. el Jefe del Estado y Generalísimo de los Ejércitos de España, Don Francisco Franco Bahamonde. Respuesta del Emmo. y Revmo. Sr. Cardenal-Arzobispo de Santiago de Compostela, Don Fernando Quiroga Palacios. *Compostela*, 31: 2–6.

Compostela, 1962. Los Amigos del Camino de Santiago. *Compostela*, 50: 23.

Compostela, 1964. El Patronato Nacional. *Compostela*, 55: 15–18.

Compostela, 1965. Nueva vida al Camino de Santiago. *Compostela*, 60: 21–2.

Corriente Cordoba, J.A., 1993. *El camino y la ciudad de Santiago de Compostela. Su protección juridica. Recopilación de normas*. Santiago de Compostela: Xunta de Galicia, Consellería de Cultura, Dirección Xeral de Promoción do Camiño de Santiago.

Domínguez García, J., 2009. St James the Moor-slayer, a new challenge to Spanish national discourse in the twenty-first century. *International Journal of Iberian Studies*, 22: 69–78.

Esteve Secall, R., 2005. Turismo y religión. El impacto económico del turismo religioso; especial referecia al turismo jacobeo. In X. Pardellas, 2005. *Turismo religioso: o Camiño de Santiago*. Vigo: Universidade de Vigo, 137–56.

Frey, N., 1998. *Pilgrim Stories: On and Off the Road to Santiago, Journeys Along an Ancient Way in Modern Spain.* Berkley: University of California Press.

Herrero, N., 2008. Reaching 'Land's End': new social practices in the pilgrimage to Santiago de Compostela. *International Journal of Iberian Studies*, 21: 131–49.

Herrero, N., Fife, W. and Roseman, S., 2008. The cultural politics of tourism and heritage in Galicia: Anthropological perspectives. *International Journal of Iberian Studies*, 21: 73–86.

Juanatey Heredia, D. and Santos Solla, X.M. 2005. Turismo e cartelaría en Galicia. In: X.M. Santos Solla, ed., 2005. *Galicia en cartel. A imaxe de Galicia na cartelaria turística.* Santiago de Compostela: Universidad de Santiago de Compostela, 13–40.

Leira Lopez, J., 2004. A peregrinación e o turismo cultural a Compostela. *Santiago Apóstolo dende a memoria, catálogo de la exposición.* Santiago de Compostela: Xunta de Galicia, Consellería de Cultura, Comunicación Social e Turismo, S.A. de Xestión do Plan Xacobeo, 51–65.

Lois González, R., 2013. The *Camino de Santiago* and its contemporary renewal: Pilgrims, tourists and territorial identities. *Culture and Religion: An Interdisciplinary Journal*, 14: 8–22.

Lois González, R. and Medina, J., 2003. Cultural tourism and urban management in North West Spain: the pilgrimage to Santiago de Compostela. *Tourism Geographies*, 5, 446–60.

Lois González, R. and Castro Fernandez, B.M., 2006. Se loger dans le passé: la récupération emblématique de l'Hostal des Rois Catholiques de Saint-Jacques de Compostelle en hôtel de luxe. *Espaces et Sociétés. Revue critique internationale de l'amenagement de l'architecture et de l'urbanisation*, 126: 159–77.

Lois González, R. and López, L., 2012. El Camino de Santiago: una aproximación a su carácter polisémico desde la geografía cultural y el turismo. *Documents d'Anàlisi Geogràfica*, 58(3): 459–79.

López Alsina, F., 1993. El Camino de Santiago como eje del desarrollo urbano en la España medieval. *ICOMOS*, 2: 50–60.

López Alsina, F. 2013. *La ciudad de Santiago en la Alta Edad Media.* Santiago de Compostela: Universidad de Santiago de Compostela and Consorcio de Santiago. 2nd ed.

Lopez Palomeque, F., 2000. El Patrimonio y las nuevas formas de turismo: ciudades y villas camineras jacobeas. In L. Lopez Trigal, ed., 2000. *Ciudades y Villas Camineras Jacobeas.* León: Universidad de León, 51–69.

Márquez Villanueva, F., 2004. *Santiago, trayectoria de un mito.* Barcelona: Bellaterra.

Mouriño Lopez, E., 2001. *Activación patrimonial e ritualizació*n no proceso de revitalización *do Camiño de Santiago.* Santiago de Compostela: Universidad de Santiago de Compostela.

Murray, M. and Graham, B., 1997. Exploring the dialectics of route-based tourism: the Camino de Santiago. *Tourism Management*, 8, 513–24.

Otero Suarez, J., 1999. A tumba de Santiago, entre a fe e a arqueoloxía. In *Compostela na historia: redescubrimento-rexurdimento*. Santiago de Compostela: Xunta de Galicia, Consellería de Cultura, Comunicación Social e Turismo, 15–22.

Precedo Lafuente, J., 1999. Anos Santos composteláns entre 1931 y 1975. In *Compostela na historia: redescubrimento-rexurdimento*. Santiago de Compostela: Xunta de Galicia, Consellería de Cultura, Comunicación Social e Turismo, 59–70.

Rodriguez, M.F., 2004. *Los Años Santos compostelanos del siglo XX. Crónica de un renacimiento*. Santiago de Compostela: Xunta de Galicia, Consellería de Cultura, Comunicación Social e Turismo.

Santos Solla, J.M., 1993. El Camino de Santiago como alternativa turística de Galicia. In M.P. Torres Luna, A. Perez Alberti and R.C. Lois González, eds, 1993. *Los Caminos de Santiago y el territorio. Congreso Internacional de Geografía*. Santiago de Compostela: Xunta de Galicia, Consellería de Relacións Institucionais e Portavoz do Goberno, 783–800.

Santos Solla, J.M., 2006. El Camino de Santiago: turistas y peregrinos hacia Compostela. *Cuadernos de Turismo*, 18: 135–50.

Singul, F., 1999. *Historia cultural do Camiño de Santiago*. Vigo: Galaxia.

Villares Paz, R., 2004. A segunda invención de Santiago. *Grial*, 161: 12–17.

SECTION III
Sacred Journeys to Home, Family and Nation

Chapter 9
Sacred Rootedness – Settling into Mobility in the Nineteenth Century American West

Nina Vollenbröker

Introduction

Amidst the turmoil of the second world war, the French philosopher Simone Weil endeavoured to underline the significance of a particular human condition. 'To be *rooted*', she wrote in 1942, 'is perhaps the most important and least recognized need of the human soul' (cited in Malkki, 1992: 24, my emphasis). Rootedness has received positive attention in Western cultures throughout the twentieth and twenty-first centuries, especially at times of unrest and change. In some academic scholarship, rootedness has been cast as an almost sanctified state of being, intrinsically linked to heaven – to 'the divinities' and 'the holy sway of the godhead' – as well as to the earth, the essential 'serving bearer' of life (Heidegger, 1971 [2001]: 147; also: Norberg-Schulz, 1976; Relph, 1976; Frampton, 1998; Cloke and Jones, 2001). Popular culture and political policy commonly offer parallel views. Confirming that rootedness is to be regarded with reverence and emphasising what is seen as the authentic essence of place, they typically authorise the image at the centre of its metaphor: a strengthened, virtuous person anchored – like a mature tree – in a nourishing, benevolent soil capable of providing long-term foothold and sustenance.

In these academic and everyday discourses, place appears absolute, stable and permanent. Rootedness, on the other hand, emerges as fragile and potentially fleeting. Once this sacred state of being is attained, it seems to depend on the individual's continued, sedentarist inhabitation of a particular location – a birthplace, a long-familiar house, region or landscape. When settlement turns into mobility and the boundaries of this well-known, supportive site are crossed, rootedness purportedly disintegrates and harmonious, meaningful emplacement is cast as turning into 'broken and dangling roots' (Malkki, 1992: 32). In keeping, a person who chooses mobility over settlement is often portrayed as disengaged or even delinquent on account of having lost 'their place' and relational agency to their roaming practice (Cresswell, 1996; Hannam *et al.*, 2004).

Rootedness, it appears, comes with the ever-present threat of its loss. But is this highly valued, venerable state of being really so delicate that it will necessarily be crushed by the unpredictability and upheaval of movement through space? Or is it possible that sacred roots could actually be founded on mobile routes (Clifford,

1997)? And what can a state of rootedness borne not out of a *static* but out of a *dynamic and fluid* context reveal about place and place-making – what can it offer to spatial discourse?

This chapter uses manuscript diaries and diverse personal artefacts (especially quilts) left behind by migrants and cowboys journeying in the nineteenth century trans-Mississippi West to illuminate the complex interrelation between rootedness and mobility, between accepted sacredness and spirited resistance.[1] It argues that the mobile men and women did not become up-rooted and spatially detached as they departed the familiar environment of their previous home-place for an extended period of mobility. On the contrary, their texts and objects indicate the itinerant Westerners continued to meaningfully engage with their surroundings – the unfamiliar trans-Mississippi wilderness through which they passed – constantly making sense of the changing landscape and thus actively creating it as rich and complex places. As they emplaced themselves through these processes of mobile place-making, the migrants confirm the importance of rootedness. However, they demonstrate that this sacred state of being is not always static, pure and fragile but frequently messy, contested and gradually acquired through interaction, shift and change.

Mapping Western Lands

Amongst the many treasures that can be found in the storage vaults of the Henry E. Huntington Library in California is an extensive collection of personal diaries which, a century and a half ago, accompanied their individual owners as they covered hundreds of miles over hundreds of days on their migration to California, Oregon or Utah.

Some of these Overland Diaries have been well preserved or painstakingly repaired. They are composed of fragile paper bound in scratched leather. Others are tattered and faded, their pages stained, half burned or rodent-eaten. A few consist of loose sheets filled with the dense, sky-blue swirls of nineteenth century ink. The spelling inside most is idiosyncratic, the handwriting at times undecipherable.[2] All

---

1   This chapter draws insights from manuscript diaries written on the Overland Trails (consulted mainly at the Henry E. Huntington Library, San Marino, California, at the Beinecke Rare Book and Manuscript Library, New Haven, Connecticut and at Utah State University, Logan, Utah), from images of quilts made by emigrant women (consulted mainly in Mary Bywater Cross's *Treasures in the Trunk* (Cross, 1993) and *Quilts and Women of the Mormon Migration* (Cross, 1996)), and from cowboy songs and poetry (consulted at the Fife Folklore Archive, Utah State University Special Collections and at the Library of Congress in Washington, D.C.).

2   This chapter leaves all passages quoted as they appear in the manuscripts; unusual words and compositions have not been edited in citations.

smell of mildew and dust. All are beautiful. All tell their own, personal version of a shared story.

They commence with entries describing the often painful departure from a place of origin – a familiar house and community – before moving on to speak about the early days and weeks on the Overland Trails. The land surrounding the Mississippi River was not dissimilar to the regions the mobile diarists had left behind and many voice feelings of being 'quite at home' (Bailey, 1852) on the road. The Plains, which most migrants reached after a couple of months, were less familiar. Indiana-to-California emigrant Addison Moses Crane confesses that he has 'but a very faint idea of what is called the plains' and likens their undulating vastness to 'the ocean with its heavy billows' (Crane, 1852). This stage of the journey also started to bring perils: violent storms became common, water and food scarce. The 'strange places' of the far-Western deserts and mountain ranges, finally, appeared alien and hostile to most migrants (Warner, 1853). They were treacherous to traverse and travellers commonly lost members of their party on this last leg of their journey (Hoover, 1849–50; Colt, 1862); a few journals abruptly stop as their writers passed away on the trail (Callison, 1852), while many others depict a gruelling landscape littered with dead animals and cherished personal possessions left behind by emigrants in desperate attempts to lighten their oxen's load and arrive at their destination alive (Gorgas, 1850).

The manuscript diaries clearly highlight the changing scenery and the complex, arduous experience of the Overland passage. They underline the journeys' different stages and the distinct challenges and tragic losses these brought for the travellers. But while historians traditionally interpreted these texts mostly as stories of personal defeat and estrangement (Schlissel, 1982; Myres, 1982 [1993]), they do, in fact, reveal a more multifaceted relationship between traveller, mobility, belonging and landscape (Hallgarth, 1989). The journals, together with other texts and objects surviving from the migratory West, demonstrate how closely the travellers knew their surroundings and how embedded or 'emplaced' they were in the land. This chapter borrows the term 'emplacement' from anthropologists Brigitte Bönisch-Brednich and Catherine Trundle who define it as 'a perspective in which the subject is inextricably situated in ... a *place*' (Bönisch-Brednich and Trundle, 2010: 4). The following will demonstrate *firstly* how the Western men and women's mobility enabled the travellers to inextricably situate themselves in what was essentially an unfamiliar and shifting landscape, and *secondly* how this mobile rootedness created the West as a rich, dynamic place.

The area beyond the Mississippi was vast. To contemplate the almost incomprehensible scale of the land, 'this great wilderness of Prairie which stretches in all most all directions beyond the field of wisdom' (Clyman, 1844–46) and of their task, 'this vast and fatiguing journey on the plains' (Handsaker, 1853), many migrants actively employed their diary. In order to grasp the *space of their venture*, in other words, diarists used the *space of the page* on which they recorded their daily thoughts. Some travellers sketched out maps of the terrain and drew up sections through mountains and lakes (Frush, 1850–1852; see Figure 9.1). Others

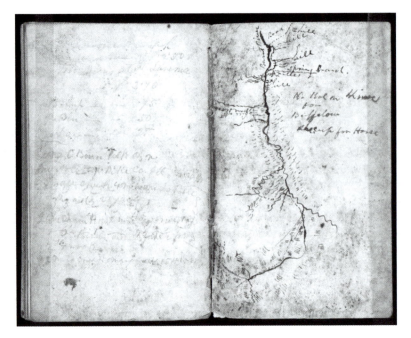

Figure 9.1   Map of the Platte River in the diary of overland traveller, William Frush

Source: Western Americana Collection, The Beinecke Rare Book and Manuscript Library, Yale University

painted the various rivers alongside which they travelled for a sizable part of their journey (Call, 1886). In addition to illustrations, diarists also used their written text, especially its position on the journal's page, to help locate themselves in the trans-Mississippi landscape. Some migrants, for example, routinely began their daily writings by marking not only the date but by recording a distinct set of additional, unchanging information. Addison Moses Crane usually commenced his entry with a record of how long he had been travelling:

> Wednesday June 9; 39th day …
> Friday June 25th; 52nd day …
> (Crane, 1852)

Indiana to Oregon traveller George Belshaw regularly used the first lines of his daily record to log not the passage of *time* but the passage of *space* with great precision:

> 4 May. Drove 16 miles on high prairie …
> 5 May, Drove 16 miles passed one or two houses …

6 May. Drove 16 miles over prairie …
(Belshaw, 1853)

Other Overlanders consistently began their daily text with a record of the weather which was as changeable as the landscape through which they journeyed:

Fryday [sic] 20th this is a very fine day …
Saturday 21 this is a bright warm morning …
Sunday 22 this is a rainy morning …
(Washburn, 1853)

These diarists reserved the very first words of their daily entries for recurring observations about their surroundings and about their own position within them. Other writers regularly returned to different special places in their books to gradually build up an understanding of their environment. A number of Overland diaries have recurring markings in their margins or dedicated mileage schedules in the back of their books (McBride, 1850; Newcomb, 1850–51). Others structured their entries with detailed records of meteorological data taken at different times of each day:

Aug 30. Plenty of frost and ice. Sunrise clear with a light breeze from the NW. Temp 30°. [morning, first entry] … Clear, light wind NW. Temp 80° [midday] … Sunset clear, light variable wind, Temp 58° [evening]. (Bruff, 1850)

Through a series of systematic entries in distinct places, the individual diaries gradually created personal maps of the complex Western space through which their owners travelled. This recurring log of geographical and meteorological data is complimented by more spontaneous information in individual entries, most obviously the invention of place names. Numerous localities in what rangeland poet Sharlot Hall identifies as 'a land without a name' were christened to reflect their unique identity (Hall, 1911: 18). Names given to particular spots by travellers often reflected local, place-specific elements and included 'Mirror Lake' (Call, 1886), 'Clinker Valley' (Love, 1853), 'Elk Springs' (Wilkinson, 1859) and 'Iron Hill':

In the morning we ascended a steep hill which I named Iron Hill
because the stones resembled iron in color.
(Longworth, 1853)

The personal journal has been interpreted as a repository for storing events or spaces a writer does not wish to engage with at the time (Culley, 1985: 8–9). Nineteenth-century Western diaries in particular have been portrayed as 'a coping mechanism', and as 'mediators … between the familiar and the new' employed to shield the travellers from the realities of their surroundings (Davis, 1987: 5). It seems, however, that quite the opposite was the case. By collapsing the seemingly endless time of the journey into the distinct and manageable units of a single day, by cataloguing the vast expanse and unpredictable weather conditions of the West into neat lists and by naming locations to reflect their distinct character, the diaries allowed the travellers to gain a distinct understanding of their surroundings and to situate themselves amidst them – a practice Bönisch-Brednich and Trundle (2010) confirm as the central part of the emplacement process. Rather than distancing them from their surroundings, the practice of daily diary writing thus opened up the Western landscape and thus allowed the diarists to place and ground themselves within it.

Becoming a Part of Western Lands

Many migrants thus used their diaries to map out a comprehensible landscape into which they could then meaningfully insert themselves. This explains why diary writing itself became an almost sacred activity to which travellers often dedicated their only free moments and which many kept up under increasingly difficult circumstances (Gorgas, 1850; Crane, 1852). But while they were working towards understanding how they formed a part of the trans-Mississippi West, the travellers were, of course, already immersed in it. The West stuck to their skin in the form of humid heat and thick desert dust: 'the thermometer being one hundred and four ... the intolerable heat and dust, together with the fatigue, makes me almost sick at heart' (Hanna, 1852). It permeated their clothes in the form of ferocious thunderstorms, pounding rain and destructive winds: 'on our body cloths drenched with continually falling showers' (Hillyer, 1849). It even killed them in the form of unclean water and extreme temperatures: 'The two sick men died of cholera last night at the Mt. W. Co's camp' (Stuart, 1849–1853).

The immediate physical reality of the West constantly touched and penetrated the itinerant men and women's bodies. But while it was frequently oppressive, the landscape was also experienced as enchanting, generous and supportive. Virtually all diaries make reference to substantial dietary staples caught on the prairies and plains: to antelope, rabbits, buffalo, fish, prairie chickens and ducks. Mary Burrell wrote in 1854: '21 [July] ... Foster and man shot several ducks and divided with us ... Camped in a perfect meadow. Intended to have a concert but it turned into a supper of ducks' (Burrell, 1854). The migrants were literally eating their surroundings, and often the accounts of their lifesaving prairie harvests are embedded in a description of the land as pleasant and gentle. Instead of logging the acquisition of essential provisions, the journeying men and women delighted in the landscape from which their supplies came. Mary Burrell speaks about the 'perfect meadow' in which she had her feast of roasted duck. Similarly, Pennsylvania-to-Oregon emigrant Esther Hanna incorporates her account of gathering wild fruit amidst a description of the 'beautiful' landscape: 'Friday 9 [July] Came to Hains Fork of Green River ... it is one of the most enchanting spots I have seen. This is a most lovely stream with gravelly bed, beautiful trees, flowers, and here we got our first strawberries … ' (Hanna, 1852). Celinda Hines writes:

> 15th friday [sic] [April] … As we came into the latter country we became struck with its beauty and loveliness. Never before had our eyes beheld so delightful a scene. The [illegible] rolling prairie stretching for many miles there covered with fruit trees of almost every variety – currant, gooseberry [illegible] together with some bushes of strawberry. (Hines, 1853)

In the fertile regions of the American Midwest, the Overlanders were *physically* sustained by the wholesome offerings they found. But they were also *mentally* nourished by a landscape most migrants experienced as delightful and benevolent. The Westering men and women's receptivity towards the landscape demonstrates their profound engagement with and their sensuous immersion in their surroundings. This particular way of rooting oneself has been considered by numerous academic scholars (although, of course, in a different context). Clifford Geertz (1996: 260) describes it as 'person and place pour[ing] into each other'. Edward Casey (1996: 19) reflects on this cherished harmony between individual and environment by offering the concept of being 'not only *in* places but *of* them'.

Geertz and Casey's notion of person and landscape becoming one can be detected in the above-cited travellers' notes about the food provided by and eaten amidst their generous and pleasant surroundings. But the itinerant men and women opened themselves up to their surroundings not just through their sense of taste. They also allowed their bodies to merge with the Western landscape's sounds and scents. In mid-February, Virginia-born James Clyman (1844–6) detected 'several of the early summer birds … chirping in the thickets'. Fellow emigrants likewise refer to the tweeting as music, 'the melody of birds' (Crane, 1852), while James Clyman himself perceived a rushing waterfall as a slow song: '12 [Sept] about sunrise we woke up again on the trail and passed the falls whose musick luled [sic.] us to sleep last night' (Clyman, 1844–6). Other travellers on the Overland Trail and on the open western range similarly welcomed the comforting 'liquid notes' of mocking birds and whippoorwills (Chittenden, 1897: 49; Love, 1853), listened to 'sand storms make music' (Lomax, n.d. 5662/JL416), to 'the whispering voice of the night wind' (Kiskaddon, 1924) and the 'songs of summer rain' (Chittenden, 1897: 69).

The mobile men and women were also acutely aware of the manifold scents that entered their bodies as they travelled. Rangeland poet Sharlot Hall records detecting the distinct 'smell of drought' as well as the 'smell of rain' (Bennett, 2001: 46) saturating the deserts and mountains surrounding her. 'Many flowers fill the atmosphere with a very pleasant odor', writes Overlander Mary Burrell (1854) and fellow emigrant Esther Hanna (1852) confirms 'the air [is] heavy with the odor [of wild roses]'.

For migrants and cowboys, these ephemeral sensuous elements frequently became physical. In their texts, the morning desert air and the warmth of the sunshine can be drunk like wine (Pettey, 1910: 33) and their skin is 'caressed' by the west wind 'blowing dewy fragrances in the morn' (Chittenden, 1897: 6–7). Rangeland poet Wesley Beggs (1912: 29) felt so strongly connected with the land that he describes his being on the plains as 'Nature just a-hugging me'. Fellow writers refer to 'night [drawing] her curtains around us' (Chittenden, 1897: 65) and speak about being 'wrap[ped] ... up in [the] ... caresses' of the Western trails (Brubacher, n.d.: 60). Possibly most explicit about achieving a state of sacred rootedness through becoming part of the surrounding landscape is this roaming poet who writes:

> The cow man's eye is made for seein' distance,
> Something out there where the blue mists roll,
> He's become a part of all these spaces,
> And all the spaces part of his soul.
> (Brubacher, n.d.: 9)

While the migrants used their diaries to map the vast and shifting trans-Mississippi landscape and contextualise themselves within it, their mobility was very much an embodied experience and most itinerant men and women received the landscape with all their senses. The migrants allowed Western space to touch them (Maddrell and della Dora, 2013: 1106), both physically, as this chapter has already shown, and spiritually, as it will now go on to explain.

Caring for Western Lands

The pages of Celinda Hines' (Hines, 1853) and William Findley's (Findley 1845) diaries are not just marked by the travellers' handwriting; they also bear the imprint of flowers both picked from the prairie and carefully preserved on their journey to Oregon. These dried blossoms recall other mementos commonly found in travel journals everywhere: the 'relics' of a departed home. Western travellers carried a variety of these revered keepsakes in their diaries: the photo of a wife who had stayed behind in Missouri (Willis, 1849), the lock of hair from a daughter or the calling card received from a now-distant friend (Warner, 1853). If those items can be seen as tokens taken on a journey because they held the sacred memory of a cherished place left behind (Hannam *et al.*, 2004), then Celinda Hines' and William Findley's collecting from the Western landscape may also be interpreted as the wish to keep with them a piece of a venerated site for which they felt affection, love and care.

The attitude of affection and caretaking shown by some Western travellers recalls the work of Martin Heidegger as well as a number of phenomenological thinkers inspired by his thoughts (Heidegger, 1971 [2001]; Relph, 1976; Seamon and Mugerauer, 1989; Feld and Basso, 1996). Heidegger identifies an ethics of care towards their unique surroundings as an indicator of a person's rootedness. Preserving and 'taking care of something in intimate familiarity', he argues, allow for dwelling, for being in the world (Heidegger, c.1979 [1992]: 158). This position of spatial understanding and responsiveness, of environmental preservation and caretaking is omnipresent in the nineteenth century travellers' texts, artefacts and practices.

Acts of tender collecting and conservation like Celinda Hines' and William Findley's abounded on the Overland Trails. Travel diaries confirm that the westward-journeying men and women took flowers and tree bark (Willis, 1849), petrified butterflies (Crane, 1852), birds' feathers, shells and pieces of flint and quartz (Call, 1886) into their custody, safeguarding them in special spots amongst their luggage. One traveller attempted to gather seeds so he might then procreate inter-mountain flowers upon his arrival in California (Mann, 1849). What is more, a number of women collected from the environment of their travels in an even more personal and caring way. Giving birth on the Trail, they named the new family members Humboldt, Sierra Nevada or Idaho Montana, forever keeping with them a fragment of the unfamiliar Western sites that first welcomed their babies into the world (Myres, 1982 [1993]).

The travellers commonly and liberally received other elements of the mobile West into their circle of close friends or into their sacred family lore, too. Many emigrants, for instance, rejoiced upon sight of the Platte River along which they travelled for a substantial part of their journey, referring to it as 'our friend the Platte' in their diaries (Heywood, 1850–56). Other Westerners liken trees to 'old friends' (Ashley, 1852) and the wind and sun to close allies (Lomax, n.d.: 5653/JL222). Rangeland writer Katherine Fall Pettey (1910) casts cacti as ever-watchful sentinels welcoming the roaming cowboy intimately familiar with his desert surroundings and fellow cowboy poet Sharlot Hall (1922: 20) speaks about the far Western 'storm-swept clouds' and stars as 'my kind and kin', highlighting her spiritual connection with these elements.

In addition to their words, drawings and collected mementos, another set of personal artefacts speaks about the state of groundedness the Westerners found in their mobile practice: the patchwork quilts made by the travelling women. Following a number of 1970s and 80s feminist publications demonstrating the historic relegation of female artistic talent and deconstructing the patriarchal art/craft binary (Parker and Pollock, 1981; Nochlin, 1989), needlework is now taken seriously as a site of conscious personal expression (Showalter, 1994; Cross, 1996; Feyerabend, 2010), but the quilts of the nineteenth-century American migration have received limited scholarly attention (with the exception of Cross, 1993 and 1996).

A considerable number of quilts survive in museums, historical societies and private collections today. Possibly the most popular quilt pattern of mid to late eighteen hundreds was the 'Log Cabin', a composition which references a type of dwelling commonly constructed in timbered regions of the frontier West and represents this structure in its design (Showalter, 1994: 153). A 'Log Cabin' quilt is composed from individual panels that each have a distinct centre surrounded by a thick border with square-notched corners. The border denotes the heavy walls which typically enclosed this crude residential structure, the focal patch denotes the interior space and central hearth of the house. Log Cabin patterns, quilt historian Ruth Finley (1970: 28) points out, 'memorialize the settlement of a continent' and it might therefore be expected that migratory women would have chosen them as templates for their own designs – after all, they were on the move to make a new home for themselves (and for their nation) in the West. So was this the case? Did mobile women chose to use the voice that their needlework gave them to commemorate their quest for a the stable home symbolised by the solid logs and warm hearth of the log cabin?

The surviving quilts indicate that needleworkers who were journeying, or had recently been journeying, through the nineteenth century American West did occasionally assemble compositions featuring this sedentary design. A far greater number of mobile quilters, however, created different compositions: instead of a 'Log Cabin' they elected to make a 'Birds in Flight', a 'Pin Wheels', a 'Road to California' or a 'Wandering Foot'. Given the fact that quilts are probably the only form of craft where each item is always given its own name and that this name is an essential part of the finished piece (Lithgow, 1974: 54), the quilters' conscious choice of titles which emphasise movement rather than settlement is telling. What is more, it was not only their names but also the iconography of these quilts that celebrated their makers' journey across the continent. One young migrant, gradually assembling a quilt as she slowly moved West in her covered wagon, carefully inked the different stages of her passage onto her cloth patches. 'Left Hamilton, April 15, 1859' the squares of fabric read, 'Crossed the Plains' then 'Seven months on the road' and 'Arrived in Columbia, Oct 28. 1859 (Laury, 1990: 43)'. In a similar vein, another mobile needleworker commemorated her fellow Overland travellers by recording the names of most men and women travelling in her 1859 train on the quilt she was stitching together while moving West alongside them (Ferraro *et al.*, 1987). These practices (the choice of mobile patterns and quilt names, the recording of routes and paths, the commemoration of fellow migrants) clearly foreground movement, suggesting that the *traversing*, not *settling* the West was what the needleworkers wanted to memorialise. The quilts can therefore be seen as writing a story of movement into the travelling individuals' personal history; they served to sacralise mobility, not settlement, in family lore.

Creating Western Lands

It is clear that the itinerant men and women in the trans-Mississippi West continually engaged with the vast and shifting landscape that surrounded them. They observed it closely and used their cartographical and meteorological studies to understand their position within it. They immersed themselves in its scents, sounds and tastes. They extended an attitude of affection and care towards it. They committed their movement to family history. Through these complex, ongoing practices the mobile Westerners constantly strove to achieve the sacred state of rootedness they no doubt aspired to.

The nineteenth century Western journeys thus suggest that rootedness is not a once-bestowed, static position which is pure, fragile and absolute. Instead, they reveal rootedness as a multi-layered and constantly developing, shifting process which is contested and fought for. And if rootedness can be seen as evolving, as not static but essentially mobile itself, then it is no surprise that this sacred state of being is supported by physical mobility, by human movement through space. This chapter therefore argues that the trans-Mississippi migrants achieved a meaningful relationship with their surroundings *because* of their mobility and the ongoing spatial interrogation and engagement that came with it, not despite it. The question which thus remains is: what can this understanding of rootedness as contested and mobile reveal about place? How did the nineteenth-century travellers' continual process of mobile rooting in turn produce the West? The mobile quilts, again, offer insights.

Comparing them to architects' drawings, Dolores Hayden and Peter Marris imply that quilts depict space at a particular scale; that of a door handle, a house or a residential plot (Hayden and Marris, 1981: 39–40). However, the question of scale in patchwork designs is a tricky point. Quilt patterns are ambiguous in terms of size and nineteenth century needleworkers would have proportionally translated the outlined shapes of a design to the desired dimensions, thus creating a composition to suit their home or the fabric scraps they had available. A pattern is therefore never truly 'to scale' or 'out of scale'. Still, the Trail quilts seem to generally share the same distinct scale. This is not the standard 1:5 for a detail or 1:50 for a plan or section – architectural scales which observe how the human body relates to the immediate environment it occupies. The mobile quilts' scale is much larger, as becomes clear not just from the completed items but also from the processes of their production.

Elaine Hedges (1987: 18) explains how the lengthy process of assembling a quilt was commonly broken down into a multitude of shorter sewing sessions: 'The method usually employed ... was the *stint* – the assignment of a specific amount of work to be done each day'. The majority of nineteenth century quilters were mothers whose days were split into disjointed patches of time and in the sedentary home, large sewing projects were divided into 'stints' and fitted around domestic chores. On the Trails, quilting was similarly organised around the everyday necessities of mobile life – looking after children, gathering food or fuel, cooking, driving the oxen – and most needleworkers honoured the stint, stitching together only a few fabric patches every day. Doing so, a mobile quilter translated geographical space into a two-dimensional representation. Every time she paused in her day to sew a patch, she marked a particular point within her quilt (a half-square triangle) and within her surroundings (a camp by the Little Blue River). When she returned to her sewing project, the quilter repeated this process, placing another patch next to the work the previous day's stint had produced and stitching the two together. Through her actions, however, the quilter did not only bring together the two fragments of cloth. She also brought together the experience of the geographical location where the first patch was made (a camp by the Little Blue River) and the location where the second patch was made a day's journey further west.

The mobile needleworker's journey and her quilt progressed parallel to each other and the traveller necessarily translated the increments of her trip onto her work. The scale of a Trail quilt, then, is not one of architectural detail. Instead, it is the enormous scale of slowly-traversed distance, the many hundreds of mobile miles translated into cloth. This is possibly best illustrated by one migrant's embroidered inscription on the back of her quilt: 'Elizabeth Jane Rogers Shepherd; Pioneer of 1866; 10,513 pieces' (Figure 9.2).

**Figure 9.2   'Trip Around the World' quilt made by Elizabeth Shepherd
            (quilt size: 213 cm × 198 cm)**

Source: International Society Daughters of Utah Pioneers

The quilts of the mobile West thus map out an intimate engagement with much larger space than is usually associated with rootedness. This larger-scaled emplacement also surfaces in the travel diaries and in rangeland songs. When the Overlanders settled at night, for example, they frequently spatialised valleys or a stretch of prairie as a domestic home. They write about sleeping 'on the bed of new grass' (Crane, 1852), 'with a rock for a pillow' (Longworth, 1853) and 'under the broad canopy of heaven' (Pengra, 1853) which arches over them like a ceiling. Cowboy poets similarly describe the wide open Western landscape using a vocabulary more frequently associated with smaller, even domestic spaces. Again, the skies become a protecting roof (Beggs, 1912), stars are seen as ceiling lamps (Beggs, 1912), the grassy prairies as a soft carpet or as a comfortable bed (Lomax, n.d.: 5654/JL305), and the sheer expanse of the surrounding land becomes an enclosing boundary wall (Brubacher, n.d.: 59). Rangeland poet Sharlot Hall analogises:

> Sky for Roof and the earth for floors;
> Home as big as all outdoors.
> (Hall, 1911: 219)

The scale in which many of the itinerant men and women felt comfortable is larger than the scale commonly associated with the human body and with human rootedness. But the travellers' written text does not merely confirm that they are indeed meaningfully positioned in a differently-dimensioned landscape. They further highlight that the itinerant men and women were enfolded in a landscape which – like the travellers themselves – continually moved and changed. They highlight that the nineteenth century migrants understood the West itself as mobile.

Virtually all mobile diarists were attentive to the unstable, shifting nature of their surroundings. In extended and often lyrical diary entries their texts closely follow sunrises and sunsets, track clouds dancing above the desert floor, watch storms forming in the distance and slowly approaching (Love, 1853; Warner, 1853). Many highlight the particular aura of certain times of day and the subtle yet distinct changes in light conditions and environmental ambiance they brought with them (Hall, 1911: 79; Crane, 1852; Hanna, 1852). Cowboy songs emphasise silent, pale grey fog continually drifting through valleys (Pettey, 1910: 53) and address the landscape's more long-term shifts, contemplating the age and formation of mountain ranges, canyons, rivers and vegetation (Pettey, 1910: 25; Hall, 1911: 43; Brubacher n.d.: 5).

The quilts of the Western migration similarly highlight place as mobile. When read as maps of the surroundings in which their authors had found a sense of emplacement, their shapes and distinct directionality create a landscape of shift and change. Blue-green pinwheel shapes set 'on point' reference wind; summer breezes, autumnal mists, wintery snow storms (Cross, 1993: 29). One-thousand brown and butterscotch-yellow triangles recall a flurry of falling leaves (Cross, 1993: 60). Footprints observe the annual migration of animals and birds across the American continent (Cross, 1993: 69). These dynamic patterns encourage the viewers' eyes to move freely across the entirety of the quilts' surface up to their – often un-framed – borders.

As the nineteenth century migrants developed a sacred emotional attachment to the West and invested it with meaning – as they *found their place* in it – they in turn *created place* (Lefebvre, 1991; Cresswell, 2004). The texts and artefacts they left behind highlight that the place through which the itinerant men and women continually produced a sense of rootedness did not only accommodate their mobile bodies but was *itself mobile*. The travellers on the Trails, frontiers and open ranges were acutely aware of the West as a landscape on-the-move and embraced it as a dynamic entity; they clearly highlighted its ongoing shifts and perpetual rhythms – be they located in weather and season, in gradual, long-term transformation or in the changing practices and inhabitations carried out by the migrants themselves. Rootedness, when seen as an active state of being, as a process, thus allows for an understanding of both people and places as dynamic, changing, travelling.

Conclusion

Rootedness is commonly perceived as sacred. It is frequently discussed in academic, political and popular discourse, especially at times when change or turmoil are perceived as threatening this supposedly pure and vulnerable state of being. This understanding portrays the rooted individual as strong or grounded and the 'ground' itself – a person's place – as stable and absolute. The loss of place, and hence of rootedness, is perceived as tragic and mobility is cast as a common cause of such a loss.

This chapter has demonstrated that mobility does not necessarily result in the loss of rootedness. Looking at nineteenth century diaries, songs and needlework, it has shown that meaningful emplacement continued to be important to the mobile men and women in the American West. Far from disengaged from their environment, migrants and cowboys commonly approached the landscape with interest, understanding and affection and related to it as a site of meaning and as a field of care.

While rootedness might have remained highly cherished, even sacred to the Western migrants, they nonetheless demonstrated that this state of being is far from being either neat and pure or fragile and attained once and for all. Instead, they reveal rootedness as an ongoing process which is often complex, messy and contested – an active participant in the construction of place rather than a receiver of place. In keeping, rooted individuals emerge not as passive keepers of this state of being but as involved, inventive and resilient men and women who are at home in mobility, shift and change.

When comprehended as ongoing process, rootedness can no longer be coerced to authorise essence-based, stable visions of neatly bounded place. Instead, meaningful emplacement depends on place which is itself dynamic and on-the-move, incessantly shifting and changing. The migrants in the nineteenth century American West beautifully demonstrate this understanding of place as a mobile, flowing landscape.

It seems fair to conclude, then, that the lengthy and challenging journeys West made by the thousands of migrant men and women can be seen not as pilgrimages *to* eventual static rootedness attained at the end of their journey but instead as pilgrimages *of* ongoing, mobile rootedness.

References

Diaries

Ashley, A.J., 1852. Crossing the Plains. [Diary]. San Marino, CA: Henry E. Huntington Library.

Bailey M.S., 1852. A *Journal of Mary Stuart Bailey, Wife of Dr Fred Bailey, From Ohio to Cal.* [Diary]. San Marino, CA: Henry E. Huntington Library.

Belshaw, G., 1853. *Diary of a Journey from Indiana to Willamette Valley* [Diary]. San Marino, CA: Henry E. Huntington Library.

Bruff, J.G., 1849–50. *Journal.* [Diary]. San Marino, CA: Henry E. Huntington Library.

Burrell, M., 1854. *Mary Burrell's Book.* [Diary]. New Haven, CT: Beinecke Rare Book and Manuscript Library.

Call, C.B., 1886. *Diary from Salt Lake City, Utah, to Los Angeles, California.* [Diary]. San Marino, CA: Henry E. Huntington Library.

Callison, J.J., 1852. *Diary of Oregon Trail.* [Diary]. San Marino, CA: Henry E. Huntington Library.

Clyman, J., 1844–46. *Travel Dairies.* [Diary]. San Marino, CA: Henry E. Huntington Library.

Colt, M.D., 1862. *Went to Kansas; Being a Thrilling Account of an Ill' fated Expedition to That Fairy Land and its Sad Results.* Watertown: L. Ingalls and Co.

Crane, A.M., 1852. *Journal of a Trip from Lafayette, Ind., to Volcano, Calif., via Fort Laramie, Salt Lake City and the Humboldt River.* [Diary]. San Marino, CA: Henry E. Huntington Library.

Findley, W., 1845–1847. *Overland journey to Oregon and a return trip.* [Diary]. New Haven, CT: Beinecke Rare Book and Manuscript Library.

Frush, W.H., 1850–1852. *Overland Journey Across the Plains to Oregon.* [Diary]. New Haven, CT: Beinecke Rare Book and Manuscript Library.

Gorgas, S.A., 1850. *Diary of an Overland Journey from St Joseph MO to Placerville CA by way of Fort Laramie.* [Diary]. San Marino, CA: Henry E. Huntington Library.

Handsaker, S., 1853. *Journal of an Overland Trip to Oregon.* [Diary]. San Marino, CA: Henry E. Huntington Library.

Hanna, E.B., 1852. *Diary of Overland Journey from Pittsburgh, PA, to OR.* [Diary]. San Marino, CA: Henry E. Huntington Library.

Heywood, M.S., 1850–56. *Journal.* [Diary]. New Haven, CT: Beinecke Rare Book and Manuscript Library.

Hillyer, E., 1849. *Overland Journey from Ohio to California.* [Diary]. New Haven, CT: Beinecke Rare Book and Manuscript Library.

Hines, C.E., 1853. *Overland Journey from Hastings, New York, to Portland, Oregon.* [Diary]. New Haven, CT: Beinecke Rare Book and Manuscript Library.

Hoover, V.A., 1849–50. *Diary of a Journey from Mobile, Ala. to Los Angeles.* [Diary]. San Marino, CA: Henry E. Huntington Library.

Longworth, B.N., 1853. *Diary of an Overland Journey from Ohio to Oregon.* [Diary]. San Marino, CA: Henry E. Huntington Library.

Love, H.M.S., 1853. *Diary of a Journey from Pennsylvania to Oregon.* [Diary]. San Marino, CA: Henry E. Huntington Library.

Mann, H.R., 1849. *The Diary of Henry Rice Mann.* [Diary]. New Haven, CT: Beinecke Rare Book and Manuscript Library.

McBride, W.S., 1850. *Journal of a Trip Overland from Goshen, Indiana to Salt Lake City.* [Diary]. San Marino, CA: Henry E. Huntington Library.

Newcomb, S., 1850–51. *Journal of an Overland trip from Darien, Wis., to California and Oregon.* [Diary]. San Marino, CA: Henry E. Huntington Library.

Pengra, C.E.S., 1853. *Diary of an Overland Journey from Illinois to Oregon via South Pass.* [Diary]. San Marino, CA: Henry E. Huntington Library.

Stuart, J.A., 1849–1853. *Notes on a Trip to California, and Life in the Mines.* [Diary]. New Haven, CT: Beinecke Rare Book and Manuscript Library.

Warner A.S., 1853. *Diary of a Journey from Pennsylvania to Oregon.* [Diary]. San Marino, CA: Henry E. Huntington Library.

Washburn, C.A.S., 1853. *Journal from Iowa to Oregon Territory.* [Diary]. San Marino, CA: Henry E. Huntington Library.

Wilkinson, J.A., 1859. *Journal of a Trip across the Plains, via Sublette's Cutoff.* [Diary]. San Marino, CA: Henry E. Huntington Library.

Willis, E.J., 1849. *Diary of Edward J. Willis. Giving Account of Travel from Independence Missouri to California in 1849, Across the Plains.* [Diary] New Haven, CT: Beinecke Rare Book and Manuscript Library.

Other Sources

Beggs, W., 1912. *Rhymes from the Rangeland.* Denver, Colorado: The Eastwood-Kircher Printing Company.

Bennett, V., ed., 2001. *Cowgirl Poetry: One Hundred Years of Ridin' and Rhymin'.* Salt Lake City, UT: Gibbs Smith.

Bönisch-Brednich, B. and Trundle, C., 2010. Introduction: Local Migrants and the Politics of Being in Place. In B. Bönisch-Brednich and C. Trundle, eds, 2010. *Local Lives: Migration and the Politics of Place.* London: Ashgate. 1–16.

Brubacher, E.A., [n.d.]. *Songs of the Saddle and Trails Into Lonesome Land.* Boise: Bess Foster Smith Publisher.

Casey, E.S., 1996. How to get from Space to Place in a Fairly Short Stretch of Time. In S. Feld and K.H. Basso, eds, 1996. *Senses of Place.* Santa Fe, NM: School of American Research Press, 13–52.

Chittenden, W.L., 1897. *Ranch Verses*. New York: The Knickerbocker Press.

Clifford, J., 1997. *Routes; Travel and Translation in the Late Twentieth Century*. Cambridge, MA: Harvard University Press.

Cloke, P. and Jones, O., 2001. Dwelling, place, and landscape: an orchard in Somerset. *Environment and Planning A*, 33: 649–66.

Cresswell, T., 1996. *In Place/Out of Place; Geography, Ideology, and Transgression*. Minneapolis, MN: London: University of Minnesota Press.

Cresswell, T., 2004. *Place: A Short Introduction*. Oxford: Blackwell.

Cross, M.B., 1993. *Treasures in the Trunk; Quilts of the Oregon Trail*. Nashville: Rutledge Hill.

Cross, M.B., 1996. *Quilts and Women of the Mormon Migrations*. Nashville: Rutledge Hill.

Culley, M., ed., 1985. *A Day at a Time: the Diary Literature of American Women from 1764 to the Present*. Old Westbury, NY: Feminist Press.

Davis, G.R., 1987. Women's frontier diaries: writing for good reason. *Women's Studies*,14(1): 5–14.

Feld S. and Basso K.H., eds, 1996. *Senses of Place*. Santa Fe, NM: School of American Research Press.

Ferraro, P., Hedges, E. and Silber, J., 1987. *Hearts and Hands; The Influence of Women and Quilts on American Society*. San Francisco: Quilt Digest Press.

Feyerabend, B., 2009. Quilting Auto/Biographies. In A. Hornung, ed., 2009. *Auto/Biography and Mediation*. Heidelberg: Winter. 89–100.

Finley, R., 1970. *Old Patchwork Quilts and the Women Who Made Them*. Newton Centre, MA: Charles T. Branford Co.

Frampton, K., 1998. Towards a *Critical Regionalism*: Six points for an architecture of resistance. In Foster, H., ed., 1998. *The Anti-Aesthetic; Essays on Postmodern Culture*. New York: New Press, 17–34.

Geertz, C., 1996. Afterword. In S. Feld and K.H. Basso, eds, 1996. *Senses of Place*, Santa Fe, NM: School of American Research Press, 259–62.

Hall, S., 1911. *Cactus and Pine; Songs of the Southwest*. Boston: Sherman, French and Company.

Hallgarth, S., 1989. Women Settlers on the Frontier: Unwed, Unreluctant, Unrepentant. *Women's Studies Quarterly*, 17(3 and 4): 23–34.

Hannam, K., Sheller, M. and Urry, J., 2004. Editorial: Mobilities, immobilities and moorings. *Mobilities*, 1(1): 1–22.

Hayden, D. and Marris P., 1981. The Quiltmaker's Landscape. *Landscape, 25(3):* 39–47.

Heidegger, M., 1971 [2001]. *Poetry, Language, Thought*. New York: Perennial Classics.

Heidegger, M., c. 1979 [1992]. *History of the Concept of Time*. Bloomington: Indiana University Press.

Kiskaddon, B., *1924. Rhymes of the Ranges. Hollywood: Earl Hays.*

Laury J.R. and California Heritage Quilt Project, 1990. *Ho for California! Pioneer Women and their Quilts*. New York: E.P. Dutton.

Lefebvre, H., 1991. *The Production of Space*. Oxford: Blackwell.

Lithgow, M., 1974. *Quiltmakers and Quiltmaking*. New York: Funk and Wagnalls.

The Lomax Papers, Fife American Collection, Folk Coll 4, No. 2, Vols 41–3 (in-text citations refer to Dossier number and poem number) Logan, UT: Utah State University Special Collections.

Maddrell, A. and della Dora, V., 2013. Crossing surfaces in search of the Holy: Landscape and liminality in contemporary Christian pilgrimage. *Environment and Planning A*, 45(5): 1105–26.

Malkki, L., 1992. National Geographic: The rooting of peoples and the territorialization of national identity among scholars and refugees. *Cultural Anthropology*, 7(1): 24–44.

Myres, S.L., 1982 [1993]. *Westering Women and the Frontier Experience, 1800–1915*. Albuquerque: University of Mexico Press. 2nd edition.

Nochlin, L., 1989. Why have there been no great women artists? In L. Nochlin, ed., 1989. *Women, Art, And Power And Other Essays*. London: Thames and Hudson, 145–78.

Norberg-Schulz, C., 1976. The Phenomenon of Place. *Architectural Quarterly*, 8(4): 3–10.

Parker R. and Pollock, G., 1981. *Old Mistresses: Women, Art and Ideology*. London: Pandorra.

Pettey K.F., 1910. *Songs From The Sage Brush*. Tucson: State Consolidated Publishing Co.

Relph, E., 1976. *Place and Placelessness*. London: Pion Limited.

Schlissel, L., 1982. *Women's Diaries of the Westward Journey*. New York: Schocken Books.

Seamon, D. and Mugerauer, R., eds, 1989. *Dwelling, Place, and Environment: Towards a Phenomenology of Person and World*. New York: Columbia University Press.

Showalter, E., 1994. *Sister's Choice: Tradition and Change in American Women's Writing*. Oxford and New York: Oxford University Press.

Chapter 10

Origination: Journeying in the Footsteps of Our Ancestors

Katy Beinart

Introduction

Tomorrow we set off for Antwerp to find the Green Cape and set sail for Southern Climes. We have been given many tasks to do on board, as well as a jigsaw puzzle. We are ensuring a taste of home with Earl Grey tea and marmite.
(Beinart and Beinart, 2009)

In December 2009, my sister and I took a container ship from Antwerp to Cape Town, retracing a journey undergone in the early 1900s by our great-grandparents, who had left behind their homes in Eastern Europe. For us, coming from a partially Jewish ancestry, the context of heritage tourism or roots journeying was a contested territory to set out into. An awareness of the potential dangers of what MacCannell (1973) refers to as 'staged authenticity' and Hewison (1987) refers to as 'bogus history' is only too evident from semi-humorous (and semi-fictional) accounts of similar searches. In *Everything is Illuminated* (Safran Foer, 2002), a young American Jew sets off in search of his ancestors in Eastern Europe only to discover that his tour guides do not quite offer what they advertised. Books such as *Jewish Ancestors: A Guide to Jewish Genealogy in Lithuania* (Aaron, 1996) and websites like JewishGen (Leeson, 1996) offer plentiful warnings of the difficulty of finding authentic records (an unintentional repeat of the guides aimed at the migrant ancestors, which warned potential emigrants of the less than well-intentioned people they might meet along the way on their travels). Categorising the Jewish Diaspora heritage tourism market, Collins-Kreiner and Olsen (2004) emphasise the consumption of heritage through an intermediary tour operator, and set out types of heritage-related tourism products on the market. However, many of these products promote an orthodoxy and connection to 'homeland' that provides a basis for the construction of a collective identity (Lev Ari and Mittelberg, 2008). If, as Lev Ari and Mittelberg (2008) argue, authenticity is both subjective and dialectically generated *differentially* depending on the background of the tourist, these tours leave a gap in the heritage tourism market for those whose background does not lead them to seek this orthodoxy or sense of connection. This is the space in which our self-organised journey began.

This chapter investigates the transformational effects of undertaking such a journey through a personalised account of a roots journey (Basu, 2004), making links between heritage tourism, pilgrimage, mobilities and performative art practice. I argue that the liminality of the journey itself provides a space for transformation, in which through performing invented rituals and actions, a personal form of remembrance and witnessing the past takes place (Turner and Turner 1978; Maddrell and della Dora, 2013). This encountering of place through the performance of 'embodied play' (Sheller and Urry, 2004: 4) recognises the overlaps and role changes between pilgrimage, tourism, and other kinds of business and of the relationship between 'hosts' and 'guests' as dynamic, fluid and mobile (Sheller and Urry, 2004). Heritage tourism literally embodies a range of connections, drawing on our mimetic, cultural or educational heritage as well as our ancestral heritage, and can be viewed as not exclusive but co-existent with other motivations for travel (McCain and Ray, 2003). Genealogical tourism also highlights the dualities within heritage: collective-individual, objective-imaginative and tangible-existential (Ramshaw, 2014), and these tensions shape the following account.

Growing up, I had been aware of our family archive, a collection of photographs, postcards and a *kist*[1] full of salt-spoiled silver and linens. My father had grown up in South Africa and told the story of coming to England by sea, and the archive evidenced a diasporic family, starting out in Lithuania and Russia and making their way to South Africa, Australia and America. Perhaps my own journey started most of all from a sense of absence; of religion, ritual and roots. My father's parents had died, he no longer practiced and my mother had a Jewish father and English mother, so the only mention of tradition and past came with the Borscht served at dinner, a few Yiddish words, and the long discarded prayer shawls, yarmukkes and silver candlesticks mothed in the *kist*. At a transitional point in my own life, it seemed that my ancestors resurfaced and demanded attention. Through discussions with my sister Rebecca we decided to apply for an artist's residency in South Africa to research our family's history, and to make the journey by cargo ship. Far from a 'reconstructed narrative' of heritage (Lev Ari and Mittelberg, 2008: 100) we were seeking perhaps more the sacred experience of the pilgrim – where 'journeying is said to bring the possibility of creating social and/or psychological transformation' (Coleman and Eade, 2004: 2).

Our ancestors emigrated after pogroms in Lithuania made life increasingly difficult, and in doing so they escaped the Holocaust, but I was also aware that their subsequent economic success in South Africa made for a troubled and complex relationship between victim and perpetrator, past and present. This was not a straightforward journey of sentimental admiration for the hard life or amazing escape act my forebears lived through. Rather it was one of 'uncomfortable reflexivity' (Maddrell, 2009, quoting Pillow, 2003): not seeking a comfortable answer, but rather trying to understand how their mobilities reached forward in

---

1   A South African name for a storage chest, derived from the Dutch.

time to my generation, what I was willing or wanted to accept of that heritage in my life, what I might need to forgive (and forget), and what their restless ghosts wanted from me.

Journeying in search of one's ancestors can be seen as an act of fictionalising both the past and present, of performing a movement through place which seeks to reconnect specific lost memories or events to specific sites. An act of the geographical imagination, of both the grounded reality of present place and the cerebral fantasy of one's past. In Saunders's (2011) study of Victorian and Edwardian autobiography, he discusses the vogue for fake memoirs, which he calls autobiographication. Spry (2001) identifies a form of auto-ethnography as the convergence of the autobiographic impulse and the ethnographic moment and Dorst (1987) as a text that a culture has produced about itself. This suggests that auto-ethnography can be both a methodology and a text of diverse interdisciplinary praxes. That this autobiographic impulse is relational, carried out through a deliberate movement through and connection to a particular place, is evidenced by recent research into contemporary pilgrimages and roots journeys.

Basu (2009: 2), writing about diasporic migrants returning to Scotland in search of their roots, describes a mutual process of coming-into-being of person and place: 'Through the acts of visiting – of walking, of searching, touching, photographing, sensing – person, place and narrative become fused, co-constitutive of a pervading sense of identity'. Maddrell and della Dora (2013: 1) argue for a 'surface framework' that takes into account the 'value of landscape and aesthetics in the pilgrim's experience, blending pilgrims' embodied travel, ritual, perceptions of the visual, material, affective, and the numinous'. This awareness of the relationship to surface, landscape and place as part of the roots journey or pilgrimage shows its value as a tool for imagination and transformation. Pilgrimages to one's roots take on aspects of the sacred, using ritual, and embodied movement through place as ways to experience and imagine connections to the past. The concept of mnemonics, where place acts as a trigger for memory, is built on by Ingold (2007: 16): 'the act of remembering *(was) itself conceived as a performance:* the text is remembered by reading it, the story by telling it, the journey by making it'.

So we conceived our journey as a performance, taking with us ingredients for our attempts at reconnection and re-enactment. These included an on-going artwork called *Starter Culture;* a bread culture made from grapes from our home, which along with equipment for making bread were carried in a suitcase used by previous generations of our family on the same route between Europe and South Africa (see Figure 10.1). *Starter Culture* operates as metaphor, as the nature of the culture changes from place to place and is repeatedly renewed with local flour and water; it is also a part of re-invented ritual, referring to a Russian and Eastern European Jewish practice called *Khlebosolny,* from the Russian Khleb-Sol, meaning bread and salt, and literally a term for hospitality (Christian and Smith, 1984). In this ritual, crossing the threshold to a new home is celebrated with bread and salt, the bread representing blessing and salt the preservation of the blessing. Our re-enactment of ritual has parallels with those of pilgrims, but also draws

Figure 10.1   Journey images
Source: Author's own photographs

on performance art tradition dating from the 1960s/70s. Allen Kaprow (1993) re-enacted everyday actions or 'happenings' as artworks, whilst instructions given to us by friend and family to carry out or repeat during the journey bore similarities to performance art group Fluxus (Friedman, 1998) and pilgrims' prayers (Coleman and Elsner, 1995).

Departure

> We arrived at the small railway station from where we would embark for Libau. 'Be careful you don't trip over the signal wires' my mother said, as we alighted from Berchik's wagon and crossed to the other side of the railway-line. I stepped over the glistening rails and wires. Hours passed; then with a mighty roar the engine bore down on us. We took our places in the coaches that had come to rest. The first mishap of the journey happened early on: the train window through which I was watching the world unfold itself came crashing down on my fingers. 'The Russian trains are as ramshackle as their empire' an elderly Jew standing next to me said. This empire tumbled into dust a few years later. (Schrire Robins, 2007: 11)

> *We travelled by train to Antwerp, via Brussels. I read Maus (Spiegelman, 1991) on the train and was transported between times, other trains and travels across Europe, people fleeing for their lives or captured and in transit to camps. At the station we were met by a disgruntled taxi driver who proceeded to issue warnings about the crew's desperation for female company, so that our arrival at port was tempered by a certain wariness. We weren't on any lists and we weren't expected. A man casually looked at our passports and waved us on to the Green Cape which was loading cargo. We had to climb a rickety staircase leaving our luggage on the dock where it was unceremoniously hoisted up by crane and dumped onto deck. The Steward, Niko, welcomed us with a little too much enthusiasm and showed us our cabin, and we breathed a sigh of relief. It had a lockable door.* (Beinart and Beinart, 2009)

We had set off with a pile of books and texts, unsure as to how these would coincide on the journey to unravel our identity myth. Bastea (2004: 12) ventures that 'going back to a place of the past may be the best way we have to take ourselves back in time'. In *Austerlitz* (Sebald, 2001), which centres on a search for lost identity, the main character explores narratives of place to re-find lost memories of self:

> And might it not be, continued Austerlitz, that we also have appointments to keep in the past, in what has gone before and is for the most part extinguished, and must go there in search of places and people who have some connection with us on the far side of time, so to speak? (Sebald, 2001: 359–60)

Wylie (2007: 175) has written of geographies of Sebald's work as 'more essentially *spectral* in that their concern is with the unsettling of places and selves as a primary and generative process'. We hoped to connect with the spectral geographies of our ancestors and touch moments they had experienced, through re-embodying their journey, but we were aware that this very spectrality could unsettle and regenerate our existing modes of being.

Culture

'Being so religious mad, we lived on black bread dipped in sugar water and dried out, which we had with tea and eggs' (Schrire Robins, 2007: 25). Not being so 'religious mad', we still had our codes of culturally appropriate behaviour tested. Both vegetarians, I wavered, but Rebecca's determination meant many meals consisted of little more than boiled carrot and potato, the Polish cook on board not ready to cater for a vegetarian. We got into many conversations about food and culture, what was important, and what reminded you of home. For a secular family, food was extremely important to us, and perhaps more so given the lack of other ritual in our lives. De Certeau (1998: 86) writes of bread as a 'memorial', a cultural symbol of escape from poverty, and of the hardships of life and work, and our *Starter Culture* contained a remnant of home that became all the more important in the unfamiliar context of the ship.

On Christmas Day, we baked bread on the ship, using our *Starter Culture* from England, and shared it with the sailors, who said it reminded them of the bread they had back at home in Poland. The night before, we had been invited to share and exchange tiny pieces of communion wafer with all the officers and crew. 'The etymology of the word communion is *moi*, and signifies change or exchange. Common, communicate, mutual and renumerate all spring from the same route. Etymologically, a *companion* is a person with whom we share bread' (Newling, 2001: 119). Sharing the wafers and our bread seemed to represent the mutual exchange and communication on the ship, and the mutation of tradition, place and time. 'We don't want the *Starter Culture* to remain the same, we are interested in how it changes; as we use and adapt traditions and ceremonies from different religions and cultures' (Beinart and Beinart, 2009).

We re-enacted the *Khlebosolny* ritual at the crossing of the equator. Waking up at dawn in the middle of the Atlantic Ocean, we ate the bread and salt and thought about the passage of both place and time. This private gesture or ritual became an important way for us to retain the link to our personal journeying, a performative language that connected past to present (Connerton, 1989: 58) and recognised the internal changes taking place. Hoffman (1998) discusses this relationship between the public and private in the need to fill a lacuna, which often begins as a personal search but becomes more public. In this sense, *Khlebosolny* is a ritual which can spread, allowing others to take on and adapt the culture to their own ritual, forming a continuum between the private and public, and sacralising the journey.

Performing Identities I: Otherness

'They travelled with crates and packages – bedclothes, odd bits of furniture, tea kettles, baskets of food, samovars, phylacteries and prayer shawls. The Russian officers, leaning against the deck-rails, laughed derisively at the bedraggled mass of humanity' (Schrire Robins, 2007: 15–16).

For *Family Photograph* (see Figure 10.1), we reconstructed family photographs taken on board ship, and dressed in the original garments belonging to grandparents and great-grandparents. Rose (2010) links the domestic and the global through the family photograph, and our family archive seemed to represent this conflict between home and diaspora, with image after image of family members on board ship, going back generations.

> *What was curious was the change in sense of self, as Rebecca felt herself filling out the enormous jacket and shoes of her Granddad Ben, and Katy felt herself taking on the airs and graces of her Great-Grandmother, Edith. The photographs themselves are a deliberately unhistorical reconstruction, a hotchpotch of times and places; the setting is a ship constructed in the 1980s; the clothing dates from the 1930s and 1950s. But perhaps this is a more honest attempt to explain our confused identities than a deliberately accurate reconstruction of the past.*
> (Beinart and Beinart, 2009)

Taking the photographs, we were not aiming for historical accuracy or 'token-isomorphism' (Handler and Saxton, 1988; Macdonald, 1997) but rather an intentional citation of our families' past, in reference to Butler's (1990) ideas of performed identities, learning the postures and performances of our ancestors through re-producing them.

The idea of the diaspora can suggest either a clinging to memories or liberating from the past, either a feeling of homelessness or an unreal sense of rootedness in a fictional homeland. 'Boyarin suggests that nostalgia is a denial of the state of sustained rediasporisation, which is the nature of Jewish history' (Garfield, 2006: 101); so why attempt to reconstruct a past which most Jewish families have for one reason or another been forced to leave behind? Garfield continues: 'diasporic subjectivity offers the contemporary world a way of understanding community without statehood or attachment to territory … the emphasis is not on where you are from, but where you are going' (2006: 103). This makes sense in the context of our journey – we are not seeking a return to the past but rather, like our ancestors, perhaps more a disentanglement from narratives of orthodoxy; acknowledging the complexities of heritage, with Jewishness as part of our identity but not all.

When his father asks him not to tell the story of a significant relationship in his youth, stating 'It has nothing to do with Hitler, or the Holocaust', Spiegelman (1991: 12) retorts, 'But Pop, it's great material. It makes everything more real – more human'. It is precisely the everyday details and handed down elements of diasporic existence and of migration that constitute the human identities of our family. So for us, it's not the archetypal traditions but rather a family recipe that allow us to reconstruct an identity. And it is the changes in these elements that capture the sense of that diaspora as temporal rather than territorial. There is no better 'place' to realise that than on board ship, a no-country non-place – a heterotopia where identities can be constructed, without actual adherence to rules or nations. As Foucault (1967) states, metaphorically a ship is the ultimate heterotopia; a space of in-between-ness and otherness that functions in non-hegemonic conditions. It is a liminal space, where the lack of usual boundaries and the seclusion from mainstream society allows for a blurring of identities and positions. The liminal space of the ship reinforced the liminal state of the pilgrimage, where the pilgrims' characteristics become ambiguous, eluding the normal network of classifications (Turner, 1969; Turner and Turner, 1978; Maddrell and della Dora, 2013). As the journey continued we began to see our roles switch between travellers re-enacting a personal pilgrimage, and artist-ethnographers, participant observers drawn into a bizarre world.

> *On the day of crossing the equator, a knock came on the door and the first mate branded us on the forehead with a potato print in red, indicating our status as novices. We were then led to the deck and 'chained' together with the other new sailors, dunked in a vat of beetroot soup, tied to the bulwark and rubbed with engine oil, mustard and ketchup, 'shaved' with a toilet seat around our necks and finally made to eat bread with jam and mustard and drink coffee with salt. This final activity had to be completed in order to gain the approval of King Neptune and to be re-born as a 'rightful person of the sea'.* (see Figure 10.1)
> (Beinart and Beinart, 2009)

The experience led us to reflect on the strange conflation of life, and art, as the performance we had taken part in was far more outlandish than anything we ourselves could have dreamed up. We began to understand that in this 'non-place' (Augé, 1995) we had to create our own rules of correct behaviour for female passengers on cargo ships (see Figure 10.2).

Performing Identities II: Gender

> Logbook, SS Romeo. 1902. The passengers on the cattle-ship The Baltic with its anti-Semitic captain, took up their quarter on the floor of the hard, dung-covered, vast and gloomy stable among the hay racks and feeding troughs. Gradually they sorted themselves out according to their sex and the laws of propriety – the women taking the farthest corner. (Schrire Robins, 2007: 16)

> *We became very aware of our gender on board ship. From before we arrived we were repeatedly warned of the Polish seamen. We were told again and again how unusual it was for two young women to travel in this way. Our performance*

THE RULES

(for young female passengers aboard a Cargo Ship)

1. Eat what you are given

2. Stay in 'passenger' role:
- Sunbathe on bridge for several hours a day
- Swim in swimming pool
- Appear in Officer's Mess at meal times
- Confine yourself to your cabin in the evenings
- Watch trashy films

3. Don't flirt with crew, but do:
- Accept tea and coffee on the bridge
- Accept copious offerings of vodka
- Accept friendly pawings from older crew members
- Allow Captain to comment on your attractiveness
- Dance with everybody on designated social occasions

4. Do not try to make cakes

5. Do not try to do any manual labour

6. Do not enter other realms: stay out of crew areas

7. Always smile politely when someone explains something technical to you even if you don't understand it

8. Always express sympathy for the Captain's workload (even when you hear him watching daytime TV)

9. Accept ritual humiliation for the 'Crossing the Line' ceremony

10. Do not send burning lanterns into the cranes

Figure 10.2   The Rules
Source: Author

Victorian Lady explored a fictitious idea of a genteel past of sea travel, where lady travellers in full dress would paint careful watercolours, perhaps a more accepted version of what a woman artist should be. (Beinart and Beinart, 2009)

In contrast, for another performance, *Dangerous Cargo,* we printed our bodies with the words Dangerous Cargo and posed on the bridge; this played on the fact that we had been told that passengers were the most dangerous cargo on board ship, as they could move. The sub-text to this was the danger we presented as women, a danger of temptation and distraction. From the beginning of the voyage, the sailors told us that it was bad luck to have women on board ship. Butler's (1990) discourse on performativity questions enacted gender roles, and by deliberately performing a variety of expected roles we tested our identities and behaviours as passengers.

While waiting to depart in Antwerp, we happened upon an exhibition which included artist Hélio Oiticica's Parangoles (made-on-the-body-cape), and it seemed a strange coincidence that the name of our ship was Green Cape. During the journey we designed a Green Cape and then made a short film of Rebecca wearing the cape and dancing to unheard music on the fore of the ship. This work took on Oiticica's (1972) philosophy of non-theatre, non-ritual, non-myth; process not display; not-nostalgic but rather concrete action. The identity became not about the photograph or document, not about the past or our history, but about our being on the journey, on the ship at that very moment; it felt liberating. We were aware of our freedom as young women doing what we had chosen to do, a freedom others in previous generations would be unlikely to enjoy. It became more and more apparent that however much we wanted this journey to be about reconnecting with our roots, it was just as much about negotiating our identities in the here and now. (Beinart and Beinart, 2009)

The liminality of the ship provided us with artistic freedom, where we could test ideas, roles and transformational acts. On reflection, this was akin to a sacred space, one where we could devote ourselves to our artistic practice, commit to ourselves as artists. This journey was a crucial turning point in both of our practices as we exchanged methods of working, began a collaboration that is still going, and released ourselves from our existing patterns of work and life. In this sense the journey perhaps connects as much to the life-changing narratives of 'Frontier Tourists' described by Noy (2004) and Laing and Crouch (2011).

Translation and Miscommunication

In the mornings the waiter would ask me if I had finished. I thought 'finish' meant fish so I would sit and wait for the fish. When we got to Madeira, I was given grapes and bananas. I ate the grapes, but I threw the bananas overboard.

They tasted awful. I did not know you had to peel them first. (Schrire Robins, 2007, 26)

Migrants, like tourists, often suffer from the mistranslation of the familiar, so it becomes unfamiliar and strange. Translation, which derives from the Latin *transferre*, meaning 'to bring across', can be seen as a metaphor for migration (Basu and Coleman, 2008). An object transferred takes on a new set of significances – or may be evoked using new objects which 'stand for' the original. Similarly, in the carrying over of language, words take on new significances. Translation can also be seen as a bringing across between generations. In the same way that the translated object may become less intelligible after multiple translations, we can question how intelligible, or authentic, our experiences of ancestral seeking may be.

> *The written language of the ship is in a multitude of words: German, Polish, Italian, English; a palimpsest of makers/users/voyagers. Original signage has aged and one language has gradually replaced another, hastily typed and pasted over. The crew speak Polish, and we write out a series of questions for them, and ask the Captain to translate for us. Somehow the question 'can you give us a family recipe?' gets confused into the Polish 'what do your family think of your job?' a question that has a very different, emotive meaning. We provoke angry, difficult, upset responses, without meaning to. By the time we work out that the word 'recipe' has been confused with the word 'receive' we have already had some difficult but interesting conversations about the hardships of their lives, one perhaps we would have avoided with a question about the culture of food.*
> (Beinart and Beinart, 2009)

The mistranslation brought to the fore the present emotional geographies we were experiencing (Bondi *et al.*, 2007) which contrasted with the spectral geographies we were seeking out. In seeking the 'inter-personal authenticity' of our family ties (Wang, 1999: 364) we stumbled into the 'Inter-Personal Authenticity' of *communitas* (Turner and Turner, 1978; Wang, 1999), where the relationship between and roles of 'hosts' and 'guests' (Sheller and Urry, 2004) on board the ship became fluid, structures broke down, and our exchanges became more honest and direct.

Arrival

We were more than pleased when our wandering had come to an end. The ship now lay peacefully in the harbour and our wonder grew as we looked at Table Mountain with its tremendous tablecloth of cloud. It was one of the most magnificent sights I had seen in my life ... as wonderful was the sight of houses on the slopes of the mountain as if they had been built into its haunches. And even more wonderful when night fell and the illumined dwellings ringed the

slopes higher and higher, curving this way, then that. It was truly a city of magic.
(Schrire Robins, 2007: 28)

*On Saturday, 9 January 2010, Table Mountain appeared, distant and hazy on the
horizon. We watched all day as it grew larger and more solid, the city eventually
becoming visible at the base of the mountain. We dropped anchor in Table Bay
and spent 20 hours waiting for a space in the harbour. On Sunday, 10 January
2010, 25 days after leaving Antwerp, the Green Cape docked in Cape Town,
and we lined up our bags, ready to disembark. As soon as the Captain allowed,
we triumphantly left the ship, skipping down the rickety steps to stand on South
African soil. We carried out our Khlebosolny ritual on the dockside, marking our
arrival. Several hours later, we were still sitting on the harbour-side, waiting
for a mythical taxi that was supposed to take us to immigration.* (Beinart and
Beinart, 2009)

Our sense of anticlimax on arrival perhaps reflects that the search for our
ancestors is ultimately unresolvable; it is the journey that is important, not unlike
the pilgrim for whom the journey is the durational and transformational element
of pilgrimage, and the arrival signifies the end of that journey. Our waiting at
the dockside also forces us to realise that a large element of mobilities is in fact
stillness, or stuckness (Cresswell, 2012).

Conclusion

The act of journeying in the footsteps of our ancestors was less a re-enactment
of their exact actions, and more an embodiment of the emotions and sensations
we imagine they may have felt and experienced. Though we didn't set out on a
religious pilgrimage, elements of the experience were personally transformational,
and had parallels with sacred journeys, similar to those identified by Basu (2009),
describing homecoming journeys that become profound emotional journeys
of personal self-discovery, and Maddrell (2009), researching the emotional
geographies of pilgrims and those making journeys of remembrance.

We can understand the heritage pilgrim as a 'restless seeker for identity'
(Bauman, 1996, cited in Coleman and Eade, 2004: 5) searching for an understanding
of self, on a transformative journey that takes place on both a private and public
level. Forty (1999: 2) puts forward Connerton's (1989) argument that 'material
objects have less significance in perpetuating memory than *embodied acts,
rituals and normative social behaviour*'. The rituals we brought on the journey,
and the performative actions we carried out, are now written into our own lives
as memorials to past generations; these memories filtered through our cultural,
environmental and educational heritages, which have formed us as much as our
genetic inheritance.

Making this trip, forming a link between our own artistic practice and the past, is perhaps a more authentic means of reconnection than the 'staged authenticity' (MacCannell, 1973) of trying to directly perpetuate lost cultural or religious practices. If narratives are internalised by recreating journeys, making origin myths out of origins, we recognise the multiple points of departure and arrival – a complex of journeying – of our family. This destabilises the idea of singular roots, suggesting rather the rhizomatic, tangled roots of belonging, and the differential, personal nature of authenticity in roots journeying (Lev Ari and Mittelberg, 2008; Wang, 1999). These do not adhere to notions of collective identity or links to a particular 'homeland' (Lev Ari and Mittelberg, 2008) but rather remain diasporic, non-hegemonic, bridging past and present.

As migrant, diasporic families, so much of our culture comes from the act of journeying, the act of adapting oneself to changing environments, negotiating language, gender roles, and cultural practices. We came to understand that whether or not they sought it out, our ancestors were leaving behind certainty, clear religious and cultural identity, for the unorthodox, the unfamiliar and the liminal place of being-in-transit. By putting ourselves in liminal space, we shifted our emotional geographies and artistic practices, marking a crossing point or threshold in our own lives.

Acknowledgements

Thanks to Rebecca Beinart, my travelling companion, fellow artist and sister. Thanks also to my parents William Beinart and Jenny Stanton, and to my ancestors whose journeying created me.

References

Aaron, S., 1996. *Jewish Ancestors: A Guide to Jewish Genealogy in Lithuania.* London: Jewish Genealogical Society of Great Britain.

Augé, M., 1995. *Non-Places: Introduction to an Anthropology of Supermodernity.* London: Verso.

Bastea, E., 2004. *Memory and Architecture.* Albuquerque, N.M: University of New Mexico Press.

Basu, P., 2004. Route metaphors of 'roots tourism' in the Scottish highland diaspora. In: *Reframing Pilgrimage. Cultures in Motion*, S. Coleman and J. Eade, eds. London: Routledge, 150–74.

Basu, P., 2009. Cairns in the landscape: Migrant stones and migrant stories in Scotland and its diaspora. In A. Árnason, N. Ellison, J. Vergunst and A. Whitehouse, eds, 2009. *Landscape Beyond Land.* Oxford: Berghahn, 116–38.

Basu, P. and Coleman, S, 2008. Migrant worlds, material cultures. *Mobilities,* 3(3), 313–30.

Bauman, Z., 1996. From pilgrim to tourist – or a short history of identity. In S. Hall and P. du Gay, eds, 1996. *Questions of Cultural Identity.* London: SAGE Publications Ltd., 18–37.

Beinart K. and Beinart, R., 2009. *Origination.* Project Blog at AN Artists Talking. Available at <http://www.a-n.co.uk/artists_talking/projects/single/520058> [Accessed 29 September 2013].

Butler, J., 1990. *Gender Trouble: Feminism and the Subversion of Identity.* London: Routledge.

de Certeau, M., 1998. *The Practice of Everyday Life. Volume 2: Living and Cooking.* Minneapolis: University of Minnesota Press.

Christian, D. and Smith, R.E.F., 1984, *Bread & Salt: a social and economic history of food and drink in Russia.* Cambridge: Cambridge University Press

Coleman, S. and Eade, J., eds, 2004. *Reframing Pilgrimage: Cultures in Motion.* London: Routledge.

Coleman, S. and Elsner, J., 1995. *Pilgrimage Past and Present: Sacred Travel and Sacred Space in the World Religions.* London: British Museum Press.

Collins-Kreiner, N. and Olsen, D., 2004. Selling diaspora: Producing and segmenting the Jewish diaspora tourism market. In T. Coles and D. J. Timothy, eds, 2004. *Tourism, Diasporas and Space.* London and New York: Routledge, 279–90.

Connerton, P., 1989. *How Societies Remember.* Cambridge: Cambridge University Press.

Cresswell T., 2012. Mobilities II: Still. *Progress in Human Geography*, 36(5), 645–53.

Davidson, J., Bondi, L. and Smith, M., eds, 2005. *Emotional Geographies.* London: Ashgate.

Dorst, J., 1987. *The Written Suburb: An American Site; An Ethnographic Dilemma.* Philadelphia: University of Pennsylvania Press.

Forty, A. and Kuchler, S., eds, 1999. *The Art of Forgetting.* Oxford: Berg.

Foucault, M., 1967 [1984]. Of other spaces: Utopias and heterotopias. *Architecture, Mouvement, Continuité*, 5, 46–9. Available online at <http://foucault.info/ documents/heteroTopia/foucault.heteroTopia.en.html> [Accessed 4 September 2011].

Friedman, K., ed., 1998. *The Fluxus Reader.* Chichester: Academy Editions.

Garfield, R., 2006. Towards a re-articulation of cultural identity; Problematizing the Jewish subject in art. *Third Text 78*, 20(1), 99–108.

Handler R. and Saxton, W., 1988. Dyssimulation: Reflexivity, narrative, and the quest for authenticity in 'living history'. *Cultural Anthropology*, 3, 242–60.

Hewison, R., 1987. *The Heritage Industry.* London: Methuen.

Hoffman, E., 1998. *Shtetl: The Life and Death of a Small Town and the World of Polish Jews.* Boston: Houghton Mifflin.

Ingold, T., 2007. *Lines.* London: Routledge.

Kaprow, A., 1993. *Essays on the Blurring of Art and Life.* Berkeley: University of California Press.

Laing, J. and Crouch, G., 2011. Frontier tourism: Retracing mythic journeys. *Annals of Tourism Research*, 38(4), 1516–34.

Leeson, D., 1996. *A Jewish Genealogic Fable*. Available at <http://www.jewishgen. org/infofiles/dlfable.html> [Accessed on 26 August 2013].

Lev Ari, L. and Mittelberg, D, 2008. Between authenticity and ethnicity: Heritage tourism and re-ethnification among diaspora Jewish youth. *Journal of Heritage Tourism*, 3(2), 79–103.

MacCannell, D., 1973. Staged authenticity: Arrangements of social space in tourist settings. *American Journal of Sociology*, 79(3), 589–603.

Macdonald, S., 1997. A people's story? Heritage, identity and authenticity. In C. Rojek and J. Urry, eds, 1997. *Touring Cultures: Transformations of Travel and Theory*. London and New York: Routledge. 155–75. (Reprinted in G. Corsane, ed., 2004. *Heritage, Galleries, Museums: A Reader*. London: Routledge. 272–90.)

Maddrell, A., 2009. A place for grief and belief: the Witness Cairn, Isle of Whithorn, Galloway, Scotland. *Social and Cultural Geography*, 10(6), 675–93.

Maddrell, A. and della Dora, V., 2013. Crossing surfaces in search of the holy: Landscape and liminality in contemporary Christian pilgrimage. *Environment and Planning A*, 45(5), 1105–26.

McCain, G. and Ray, N., 2003. Legacy tourism: The search for personal meaning in heritage travel. *Tourism Management*, 24(6), 713–17.

Newling, J., 2005. The Kitchen Table. *Writings 1995–2005*. London: SWPA Ltd.

Noy, C., 2004. This trip really changed me: Backpackers' narratives of self-change. *Annals of Tourism Research*, 31(1), 78–102.

Oiticica, H., 1972. *SYNTHESIS-PARANGOLÉ brief*. Available online at <http://www.itaucultural.org.br/aplicexternas/enciclopedia/ho/index.cfm?fuseaction=documentosandcod=588andtipo=2> [Accessed 12 July 2013].

Pillow, W., 2003. Confession, catharsis or cure? Rethinking the uses of reflexivity as methodological power in qualitative research. *Qualitative Studies in Education,* 16, 175–196.

Ramshaw, G., 2014. A Canterbury tale: imaginative genealogies and existential heritage tourism at the St Lawrence ground. *Journal of Heritage Tourism*, 9, 257–69.

Rose, G., 2010. *Doing Family Photography: The Domestic, The Public and The Politics of Sentiment*. London: Ashgate.

Safran Foer, J., 2002. *Everything Is Illuminated*. Boston: Houghton Mifflin Company.

Saunders, M., 2011. *Self-Impression: Life-writing, Autobiografiction and the Forms of Modern Literature*. Oxford: Oxford University Press.

Schrire G.R., 2007. *From Eastern Europe to South Africa: Memories of an Epic Journey 1880–1937*. Cape Town: Jacob Gitlin Library, Western Province Zionist Council.

Sebald, W.G., 2001. *Austerlitz*. New York: Random House.

Sheller, M. and Urry, J., 2004. *Tourism Mobilities: Places to Play, Places in Play.* London: Routledge.

Spiegelman, A., 1991. *Maus.* New York: Pantheon Books.

Spry, T., 2001. Performing autoethnography: An embodied methodological praxis. *Qualitative Inquiry*, 7, 706.

Turner, V., 1969. *The Ritual Process: Structure and Anti-Structure.* Chicago: Aldine Publishing Co.

Turner, V. and Turner, E., 1978. *Image and Pilgrimage in Christian Culture.* New York: Columbia University Press.

Wang, N., 1999. Rethinking authenticity in tourism experience. *Annals of Tourism Research.* 26(2), 349–70.

Wylie, J., 2007. The spectral geographies of W.G. Sebald. *Cultural Geographies*, 14, 171–88.

Chapter 11

Return to Palestine

Suha Shakkour

Introduction

To begin to comprehend the complexity of life in exile for Palestinians, it is important to first recognise that for many the desire to return is not rooted simply in a return to the homeland, but to a particular region and, specifically, to their own former houses. In a sense, these houses – of which the majority were either appropriated by Jewish-Israelis (and in some cases by internally displaced Palestinians) or destroyed in 1948 – serve as place markers in history, the moment of exile forever preserved within their walls. Thus, for their original owners, a return to them is considered a return to their 'authentic selves', that is, to their pre-exilic identities. Given this, it comes as no surprise that the keys and deeds to these houses are carefully guarded and passed down through the family along with the memories of the sights, sounds, and smells they evoke in the first generation of exiles. This first generation is comprised of an estimated 726,000 Palestinians who were displaced in the 1948 *Nakba*,[1] and later the nearly 300,000 in the 1967 *Naksa*.[2] While those who had more resources (for example, an education or financial resources) were able to exercise more choice in terms of destination, many remained in neighbouring Arab countries hoping to return to their homes when the political situation was resolved. Today, more than 60 years later, the vast majority continue to wait.

Drawing on what MacCannell (1999) has called 'sight sacralisation',[3] that is the process through which a secular site attains a significance that is comparable to a sacred site, this chapter will consider how the appropriated houses of Palestinians can be seen as sacralised sites that represent both the personal and

---

1 A term used by Palestinians to refer to the exile of 1948. Translated, it means 'the catastrophe'.

2 A term used by Palestinians to refer to the defeat of the Arabs in 1967 in the Six Day War and the occupation of the West Bank and Gaza. Translated, it means 'the setback'.

3 Although MacCannell's distinction of 'site' (as a location that is visited) and 'sight' (as a location that is worthy of being seen) is useful, in many cases the appropriated houses have been destroyed and therefore cannot be visited in the same sense that MacCannell envisions. Thus, the term 'site' is instead used to refer to both scenarios throughout. This is not to indicate that they are no longer worthy of being seen, rather that our understanding of what visiting them actually entails must be adjusted.

collective experience of exile. In addressing this, it will be divided into three main parts. The first will consider the significance of 'home' and the growing popularity of diaspora tourism. The second section will consider the relationship between 'home' and the appropriated houses for first generation Palestinians residing in London, with reference to MacCannell's theory of site sacralisation. Finally, the third section will draw together the literature review and the respondents' views in a brief discussion. As will become evident throughout this chapter, although each appropriated house may be visited only by a single individual or family, the same kind of visit is repeated again and again by Palestinians to their own former houses, evoking a collective sense of identity in exile.

Defining 'Home'

'Home' may be understood as multidimensional concept that is at once a space, a place, a connection, a way of life or of existence. It has been variously defined and widely researched with emphasis being placed on belonging, identity, and security (for example, Dovey, 1985; Jackson, 1995; Wise, 2000; Rose, 2003; Tuan, 2004; Blunt and Dowling, 2005). These themes appear regardless of the geographic location of the research being undertaken, and the age, ethnicity, or religious affiliation of the respondents. 'Home', in a sense, is a universal and yet un-definable concept. There is not sufficient space here to address all the intricacies that the research on home has provided, thus what follows is an overview of some of the most salient themes that have arisen with special attention paid to the childhood house as the initial site of many of the emotions that the word 'home' conjures.

Much as with any space we create, home is 'lived as a relationship' and as such it grows and changes alongside us (Jackson, 1995: 122). This, in effect, is where much of the difficulty in defining it originates. As Massey (1994: 119) notes, there is after all 'no single simple "authenticity" – a unique external truth of an (actual or imagined/remembered) place or home – to be used as a reference either now or in the past'. Yet, since the changes are frequently in response to the inhabitants own physical and psychological needs, home comes to be seen as a space of belonging, a private sphere in which one is protected from the outside world. It allows individuals to retreat both physically and emotionally (Tuan, 1971), and thus becomes a kind of 'preferred space … [that] provides a fixed point of reference around which the individual may personally structure his/her special identity' (Porteous, 1976: 390; see also Dovey, 1985; Somerville, 1997; Young 1997; Bunkše, 2004).

This is particularly true of the childhood house, or as Malouf (cited in Crouch, 2003: 45) refers to it, 'the grounds of our first experiences'. It is perceived as a site of safety and security, an all-encompassing universe in which one is protected. While this feeling, of course, often has more to do with the relationships formed within it than with the structure itself, it is nevertheless the house that becomes inscribed in one's memories (Jackson, 1995). As a well-known Arab poet, Abu Tammam (ninth century) writes:

Take your heart wherever it desires,
Your true love will, after all, always be your first love.
How numerous are the houses in which you will reside,
Yet, never will you cease to yearn for your first house.

Given this, it is not surprising that for many, their future homes, and crucially, their 'ideal homes', are modelled after this first house (Tucker, 1995). This does not necessarily refer to the physical features of the house (though, naturally it may) but to the feeling of security that was first nurtured there. One reason for this attachment to the first house may be the fact that children are often unable to comprehend the concept of abstract time, and so the house becomes an important anchor, 'grounding' an individual's sense of identity and binding their earliest memories and dreams to a specific place (Bunkše, 2004). As Bachelard (1992: 15) notes, 'The house we were born in is more than the embodiment of 'home', it is also the embodiment of dreams'.

Although on a larger scale, and requiring a considerable amount of imagination, the same attachment can be held for one's first neighbourhood and nation of origin (Anderson, 1991). In contrast to one's personal home(s), the construction of which is generally an individual undertaking (albeit with the assistance of family and friends), the neighbourhood and the nation are imbued with significance through the individual and collective will of their residents. Through their efforts, the homeland becomes 'the landscape ... of historical memory that offers tangible images to rootedness and grounded community' (Waetjen, cited in Blunt and Dowling, 2006: 159), and its details are studied in school, through books, and even through television programmes which can introduce the individual to parts of the nation s/he may have never visited. The strongest attachments to the homeland, however, are formed through personal and collective experiences within it. Through them, an olive grove, for example, can cease to be an olive grove and instead become the site of one's first attempt to climb a tree, one's first harvest season, or in the case of many Palestinians who were living in Galilee during the Six Day War in 1967, a site of refuge.[4] As Giuliani (1991) notes, 'if the object of attachment is the place where historical, and therefore unrepeatable, events that are salient for the individual have taken place ... then, by definition it is unique and irreplaceable' (Guiliani, 1991: 7).

As a result, when an attachment to a collective place such as a nation is severed, personal and collective memories come to hold an even greater significance,

---

4 During this time, the fear of homes either being bombed and collapsing was so great that many families lived (and slept) under their olive trees. Since olive groves are traditionally planted in neat rows, each family had a square around a tree that effectively became their home during that time.

instilling within exiles a sense of being consistently 'out of place', of being forced to move forward, while always longing to return somewhere (Said, 2000). Or as Darwish (1995), a prominent Palestinian poet, puts it in describing his grandfather's experience of exile in Lebanon:

> My grandfather died with his gaze fixed on a land imprisoned behind a fence. A land whose skin they had changed from wheat, sesame, maize, watermelons, and honeydews to tough apples. My grandfather died counting sunsets, seasons, and heartbeats on the fingers of his withered hands. He dropped like a fruit forbidden a branch to lean its age against. (Darwish, 1995: 88)

Diaspora Tourism

Given the important role that the home – in both the small (own world) and large (the nation) sense of the word – plays in the development of an individual's sense of identity, the rise in what has come to be called 'diaspora tourism' is perhaps not surprising. Diaspora tourism refers to journeys undertaken for the purpose of reinforcing or re-discovering one's ancestral roots. It builds on the idea of a 'home nation' to which one is seen to belong and the desire to reaffirm a real, or imagined, personal and collective history. Included here are journeys undertaken both by individuals residing in the home nation and those residing in the diaspora, since both of their identities have all been influenced by the factors leading to the mass emigration as well as the after effects. Within this field of study, Coles and Timothy (2004) have identified six patterns of diasporic travel and tourism. They are as follows: (a) return to the homeland for the purpose of retracing the lives of one's ancestors; (b) return to the homeland for the purpose of understanding the circumstances surrounding one's ancestors' emigration; (c) return to the homeland for the purpose of reaffirming one's identity; (d) reverse diaspora; (e) the establishment of diasporic communities as tourist sites; and (f) the rising popularity of retreats created by and for members of the diaspora.

Of these, the first two patterns may be used to discuss the journeys undertaken by Palestinians in exile to their family houses in Palestine/Israel.[5] In the first pattern noted by Coles and Timothy (2004), the traveller, motivated by the desire for a renewed sense of ethnic or national identity, seeks to forge a more personal connection with their homeland. Individuals thus endeavour to trace their own family lineage before travelling, and upon arrival they may seek out former homesteads and possible distant relations (see, for example, Basu, 2007). This desire is evident among many Palestinians who make the journey back home. The intention is not only to visit Palestine/Israel, but specifically to visit the family

---

5   While the third pattern may also be applied, it generally refers to visiting sites that have a group (ethnic/cultural) significance, but are not family specific (see, for example, Bruner, 1996; Bell, 2002).

house and in so doing to establish a connection with a life that might have been their own had the circumstances been different. As Abunimah writes in describing his first visit to his family house which had been destroyed:

> Sometimes I stand in what had been a bedroom, a kitchen, a front yard, or an alley between two houses. I try to listen for the sounds of Lifta, of the life my mother told me about, of children running home from school and women calling after them, of men returning from Jerusalem and drinking tea on shady porches. (Abunimah, 1998: 3)

By seeing the family house, the visitor thus is able to 'place' many of the memories that have been described to him or her, and in the case of the first generation to relive these memories. As will be discussed later, however, the experience is not always a pleasant one. Since the houses are also the sites of exile, visits to them may also be discussed within the context of the second pattern indicated by Coles and Timothy (2004). This includes travel for the purpose of understanding the lives of one's ancestors, the route of their emigration, and sites specific to the moment of exile (the latter is sometimes referred to as 'Dark Tourism'; see, for example, Kraut, 1982; Lennon and Foley, 2000). These may include ports and railway stations, but also, as in the case of Palestinians, houses and borders. The emphasis is on movement and spaces of transit, and the goal is to be able to better comprehend their own histories – in a sense to examine the bridge between the homeland and the hostland.

For Palestinians, while the appropriated houses may be viewed as personal family specific sites, because the exile occurred on a vast scale that included the appropriation or destruction of over 400 villages, they are also perceived as part of the collective narrative of the *Nakba*. That is, the appropriated houses are seen to represent both the family home and homeland. Therefore, in visiting them Palestinians feel connected both to their own ancestors and to all Palestinians in exile. In other words, it is simultaneously an individual and collective experience. It is this collective perception of the houses that allows them to be considered as a single 'site', albeit one that is located across several different geographic places, and contributes to their sacralisation.

Site Sacralisation

As has been indicated, MacCannell (1999) posits that secular sites can become sacralised through a series of five stages. The first, in which the site is initially recognised as having some unique characteristics, is referred as the 'naming phase'. Although this may be arranged more formally, through examination by experts in the field, it may also occur spontaneously through word-of-mouth and simple repetition. In both cases, the site is deemed worthy of special attention, of having 'aesthetic, historical, monetary, recreational, and social value', and in this way is set apart from other sites (MacCannell, 1999: 44). In the second stage, referred to as the 'framing and elevation phase', the site is put on display. This is achieved through building a structure around the site, installing lights around it, or through landscaping that set it apart, depending on the site's dimensions. In all cases, the aim is both to protect the site and reinforce its status, thus demonstrating that it is worthy of such efforts.

The third stage, referred to as 'enshrinement', occurs when the objects that frame the site themselves become worthy of attention. For smaller sites (for example, a sculpture) this may be the museum that houses it, and for larger ones (for example, a building) it may be the town/city in which it is located. In both instances, the 'frame' enters into MacCannell's (1999) first stage (naming phase) and the cycle begins again. In the fourth stage, 'mechanical reproduction', replicas of the site are made and sold. This may include miniature versions of the site, photographs, and postcards. It is during this time that the site begins attracting larger numbers of visitors who wish to retain a keepsake of both the site itself and of the emotions it evoked. The replicas allow them to do so and also encourage future visitors. The final stage, referred to as 'social reproduction', occurs when the site becomes so well-known and beloved that groups or cities begin to name themselves after it. In other words, there becomes a desire for the site, and more importantly what it stands for, to exist in more than one location.

MacCannell's approach has typically been applied to tourist and secular pilgrimage sites (including national and political sites) that are visited by hundreds of thousands of individuals. Burch (2002), for instance, discusses site sacralisation in relation to Parliament Square in London, and Löytynoja (2008) considers how MacCannell's stages may be applied to sites in Northern Europe. With reference to the current study of Palestinian appropriated houses, site sacralisation, together with the phenomenon of diaspora tourism, is particularly relevant. Considered within the context of the Palestinian experiences in 1948 and 1967, and to ongoing displacements that have taken place in recent years, it helps to explain the strong bonds with home and the homeland which displaced Palestinians maintain after decades and through multiple generations, and it highlights the important connection between the individual and collective experiences in this community.

Method

The primary source of reference for exploring the relationship between Palestinians in exile and their homeland in this chapter is a series of in-depth, semi-structured interviews with seven Palestinians residing in London. These were conducted as part of a wider study of the integration and assimilation of Christian Palestinians in Britain that included 70 respondents, both Christian and Muslim (Shakkour, 2010). Those included in this chapter are the eldest respondents, ranging in age from 69 to 86 years old at the time of interview. All respondents are first generation migrants to the UK who were born in Palestine and had resided there for at least a part of their lives. The interviews took place in respondents' homes in London and the surrounding boroughs, and they varied in length, ranging between three and five hours. They were conducted primarily in Arabic and were translated and transcribed by the researcher. Respondents were asked about their early experiences in Palestine, their childhood houses, the circumstances of their migration, visits back home, and what they missed the most. As part of the wider study, noted previously, they were also asked questions relating to language, education, marriage and relationships (including views on intermarriage) as well as family life, whether they felt 'at home' in the UK, and if they intended to return to the homeland.

Given the great diversity that exists within the Palestinian population, those residing in the UK are not easily recognisable and have often been referred to, alongside other Arabs, as an 'invisible minority' (Al-Rasheed 1991; Nagel 2001). They share many physical characteristics with other Arabs and Mediterranean ethnic groups, belong to a number of religious and denominational groups and include British citizens, permanent residents, immigrants, refugees, asylum seekers and 'illegal immigrants'. In addition to this, despite the fact that there are between 5,000 and 20,000 Palestinians residing in the UK, with the majority in London (Office for National Statistics, 2001; Matar, 2005), unlike other Arab groups, they are geographically dispersed around the city.

Thus, owing to the above, respondents for this study were located using a snowball sampling method, and respondents were asked for the contact details of other Palestinians who might also be interested in participating in this research project. Among the primary benefits of utilising this approach is the establishment of a sense of familiarity even before the interview has begun. Respondents are able to 'trace' the researcher's identity and feel confident of the legitimacy of the research. As with any other sampling method, however, there are several disadvantages associated with snowball sampling. The most significant may arguably be the possibility of including only a limited set of views since respondents who know one another may also share similar opinions. In order to counter this potential source of bias, this research project includes six different snowball networks, with the initial respondents in each case being unrelated to one another. Therefore, while respondents within each network may be connected to one another in some way, it was rare that they were related to those in other networks. Nevertheless, it should also be noted here, however, that this study does not presume that it is possible to either draw generalisations about the Palestinian population as a whole, or to fully represent the individuals as they truly are. However, as Farah notes, although 'there are theoretical and methodological problems inherent in all forms of "representations", in the utilization of life-histories and in "giving voice" to the "other" … the alternative is to unwittingly collaborate in the process of silencing' (Farah, 1999: 8).

Appropriated Houses as Individual and Collective Representations of 'Home'

For Palestinians in exile, the concept of 'home' takes on an additional significance. It is at once the family house, the hometown, and the homeland – all of which they are denied access to. Several generations have been born since 1948, yet efforts persist to ensure that the past, as both a collective and individual experience, continues to be remembered. These include recounting the details of the exile through storytelling (Sayigh, 1998; Farah, 2005); scholarly writings that focus on the memories of eldest generations (Yahya, 1999; Masalha, 2008); art and literature that are based around both the moment of exile and experience of Palestinians since 1948 (Darwish, 1995; Kanafani, 2000; Boullata, 2009); and passing down traditional ways of life, including Palestinian cuisine and folk embroidery, both within families and through wider publications (Kawar and Nasir, 1980; Skinner and Kawar, 2007; Nasser, 2008). The aim is to help the younger generations to carry on as much of the pre-1948 Palestinian identity as possible, and to encourage them to continue to pursue their right to return. As Darwish (in Said, 2000) writes:

> But I am the exile.
> Seal me with your eyes.
> Take me wherever you are –
> Take me whatever you are.
> …
> Take me as verse from my tragedy;
> Take me as a toy, a brick from the house
> So that our children will remember to return. (Said, 2000: 179)

In referring to individual houses and mementoes, Darwish evokes the personal experience of exile – and the desire of those in exile to return not only to Palestine, but their own former houses. This, however, is complicated not only by virtue of the realities of exile that prevent them from returning, but also by the fact that many of these houses were destroyed or appropriated in 1948. As Masalha (2008: 130) notes, 'Of the 418 depopulated villages, 293 (70 per cent) were totally destroyed and 90 (22 per cent) were largely destroyed' (see also Khalidi, 1992). Those that remained were later appropriated by Jewish Israelis and in some instances by Palestinians who themselves had been internally displaced and had nowhere else to go. Thus, being able to return, even to visit their family houses is often an emotionally charged experience, and one that not all the respondents in this study were willing to make.

While MacCannell's (1999) model has been used primarily to discuss singular sites that have become locations of touristic rituals, it will be argued here that sacralisation may also occur in individual sites that exist across multiple geographic locations and as such are not – and do not become – commercialised. As will be discussed, in the case of the appropriated Palestinian houses, their sacralisation derives from the special significance of each house as both the ancestral family home and the site of exile. While each house may only be visited by an individual or a single family, the same visit is repeated again and again by Palestinians to their own family houses, connecting Palestinians to their houses, to one another, and to the exile itself. In this way, they come to be perceived as a single site with each house representing both the individual and collective Palestinian experience. Yet, despite their significance, it is important to note here that seeing them again can be a difficult experience. As one respondent, Mariam,[6] who had not seen her family house since 1948 noted of her visit:

> I walked down the hill and found my house – it looked *so small!* There were huge houses around it, and our garden had gone … I stood there at the gate and I saw my house, and I burst out crying … I didn't want to go in, but I just stood at the gate, and I could see my bedroom. I could see everything. It hadn't changed much, but it looked small among all the big buildings around it. Ours was the only house there when I was young … I just cried my eyes out and walked back up the little path to the taxi. I regretted going there because I regretted losing my childhood memories of the beautiful place. To me, it had changed for the worse. I just wished I hadn't gone … But, I suppose all the world has changed anyway; it would have changed whether we were there or not there.

Similarly, another respondent, Elias, who had not seen his family home since 1948, noted:

> To see my home being lived in by others, it disturbed me … We knocked at the door and one of my relatives, who spoke Hebrew, explained who we were and asked if we could see the house. [As we walked around] it raised unhappy memories. To see the things that I had used as a child being used by another family, it was very painful. [I remember thinking] 'In fact, this house belongs to me! And the grapes and fruits my father had planted as well!'

Several of the other respondents in this study, who had since returned to Palestine/Israel for a visit, expressed the same trepidation when discussing the preparations they undertook prior to making the journey. Others, however, hoping to avoid the pain they saw their friends experiencing, have never returned, even for a visit, despite their families urging them to do so. Yousef, for example, who left to pursue his studies in the early 1960s and was unable to return after the 1967 Six-Day War, never went back, even for a visit.

> I'm the only one of my brothers and sisters who has not been to Palestine since I left. It's not a phobia, but I always wanted the good memories of Palestine to remain. The more I see these pictures, and the more I see what is happening, I

---

6   All names have been changed to protect the identities of the respondents.

feel that I do not want to go and see these places that remain in my memory, [and that are so different now].

For those who did visit, the journey itself was long awaited and perceived as a kind of rite of passage, indicating that the sacralisation of the appropriated houses begins long before the visit is actually made. In most cases, this begins with descriptions of the houses that are recounted in such intricate detail that the listener may come to imagine his/her childhood as having taken place there (see, for example, Farah, 2005; Shabaneh, 2010; Shehadeh, 2010). Over time, the scents of lemon and orange trees, and the shade of the olive trees, become so repeatedly experienced in storytelling that they haunt the listener, who has never experienced them almost as much as they haunt the teller. In this way, the individual houses, their stones, their secret hiding places, come to be seen as unique and worthy of preservation by generations of Palestinians who commit their details to memory.

It is important to note here that it is the everyday details of the houses and the lands that are most keenly remembered. Thus, in returning, respondents wanted not only to see their houses, but to relive their former lives. Nawal, for example, who had not been able to return for over 50 years, recalled of her first trip back:

> The first time I went back, I wanted to visit my house, and to sit under the tree, there was an olive tree there. But I found the house and tree gone. They had built other houses in its place ... I missed the stairs, our stairs, where I [used] to sit ... I missed my grandparents' house, [the grove] of oranges, and the flowers that my grandfather had planted there.

As will be evident from the above, the surrounding land and trees frequently have their own stories and their own significance, equal to that of the house, and without seeing them the return home is incomplete. This phenomenon has been noted within much of the research that focuses on generations of Palestinians born in exile who describe their parents'/grandparents' land in great detail (Farah, 2005). The trees and land thus frame the house and themselves, become unique elements in the memories of Palestinians, a fact that was further evidenced by the presence of either an olive or fig tree in nearly every home of the over 70 respondents interviewed as part of the wider study of Palestinians in London.

With regard to the houses themselves, in most instances the respondents in this study had also retained a number of mementoes including the keys, deeds, and any other hastily collected items that have survived. This is a common phenomenon among Palestinians, and as Said (1986 [1993], 14) notes of these items, 'we carry them about, hang them up on every new set of walls we shelter in, reflect lovingly on them'. For the respondents in this study, they symbolised the houses themselves, the exile, and the promise of return, and they were proudly displayed as both reminders and, in a sense, evidence of the family's Palestinian origins and of belonging to another place. As indicated, however, this action is not unique to the homes I visited or to Palestinians in London. It is also present within the academic literature (see Jacir, 2002; Ramadan, 2009) and reinforced in fictional writing (for example, Kanafani, 2000).

The symbol of the key in particular has come to be seen as 'the cause ... that binds Palestin[ian] refugees together' (Bshara, 2008: 11). Keys are often proudly worn as pendants on necklaces, enlarged and reproduced for Land Day ceremonies, and regularly feature in Palestinian art and poetry. In his renowned short stories, for instance, Kanafani (2000) repeatedly focuses on the key as a memorial of the house and the keeper of a past life and identity. Similarly, Darwish, in recounting an interaction between a father and son on the topic of the son returning to the homeland, writes of the key almost as an extension of the individual 'He handled his key in the way he feels his *limbs*, and calmed down' (quoted in Nassar and Rahman 2008: 91; emphasis in original). Passed down through the generations, the keys of the appropriated homes are thus traced and re-traced until their owner knows them by heart, and their new owners inherit not only the houses but also the responsibility to fulfil their parents' and grandparents' hopes of returning.

In addition to keys and other mementoes, the memories of the appropriated houses are further reinforced through the retention of same sentiments and social patterns, including hospitality and form of address, which would have been present in the original home. To maintain the Palestinian way of life in exile is to ensure that when return does become possible, the traditions and customs that were present before the *Nakba* will have been preserved and can thus be built upon. More importantly for the respondents in this study, however, these small details were an essential part of the feeling of 'home' that their houses evoked and that many respondents missed. As one respondent, Melia, noted:

> When I was a child we used to come up from school and go to my grandmother's house, and then my cousins and Aunts would come ... and we'd gather. [Living in London], that's what I miss about life there. I miss the family life, the social life, the Arabic way where people would knock on doors and come in. You would have a lot of relatives around, and the larger family and this and that ... we were all much closer together.

Evidence of a desire to retain this sense of the Palestinian household was present in every house visited as part of this study, regardless of the age or gender of the respondents. In addition being offered Palestinian food and drinks, guests are encouraged to feel at home, to feel welcome. As one respondent, Walid, stated:

> This is the Palestinian identity, to let the guest feel that he is at home. We don't need to tell him to feel at home, or in English to 'make yourself at home'! No, [because then] you feel you're not at home. You're a stranger. So we don't say that. We just welcome you, and then you have this feeling [of being at home], that [this home] is for you.

The intention was to re-create a part of Palestine in the host nation for themselves, their children and their guests. As indicated previously, in many cases this also extended to patterns of decoration in the houses in London, and many included embroidered cushions and wall coverings, framed photographs of Jerusalem and Haifa, paintings by Palestinian artists, and old maps of Palestine. These were items they had either brought with them, inherited, or actively sought out in London, and together they created what may be described as a 'tangible Palestine' in exile, a way of retaining a visual reminder of their owners' sense of identity.

> Walid: We remind our children every single day, every single minute, that we belong to that land … we do that by attitude, by mentioning things like food, traditional customs, even speaking in the local dialect language, just to remind them. We call home all the time, so that they can be in direct contact with their relatives, their grandmother and grandfather. And, of course, as you see in my home, there is something called Palestine in here. So even if I don't speak to remind them, these things in here will remind them that they come from Palestine.

With each new generation there is a renewed sense of hope that this will be the one that is able to return. As one respondent, Farah, noted:

> Now it's up to you, your generation, it's your turn! For us *khalas*,[7] what more can I do now? I'm an old, sick woman with a cane. And for this reason, it's up to your generation. Now you have to work; as I say, we worked enough. I was happy when [our mutual contact] told me about you yesterday.

What may be seen here is that this sense of duty to ensure that a Palestinian way of life survives, extends beyond the respondents' own children to all Palestinians. It is, in fact, a sentiment that was repeated again and again in each of the respondents' homes, and it has been noted among Palestinians around the world (Hammer, 2005; Farah, 2007; Shabaneh, 2010). In exile, there is a heightened need to maintain some element of the homeland, to recreate a part of it on the outside so that the hope of return does not fade.

As may be evident from the above, in the same way that the visits to the appropriated houses are perceived as both individual and collective experiences, so too are the steps taken to preserve them in memory, albeit without the deliberate intention of elevating the status of the houses or encouraging a sense of unity. Rather, these were individual responses to the loss of a home and a homeland which, over time, came to hold a collective significance. For the respondents in this study, regardless of their age or gender, the appropriated houses are 'home' in every multi-faceted sense of the word. They are, at once, structures that are intended to protect their inhabitants from physical harm, spaces of security where

---

7   'Enough' is the closest translation.

one's identity need not be explained, and spaces of belonging, in both the familial and national sense of word.

In other words, for many Palestinians, including the respondents in this study, the appropriated houses represent their former and in a sense, true selves – or more accurately, who they might have become. It is for this reason that the houses are recalled and recounted in minute detail. They are sites of the moment of exile, and to return to them is a way of reclaiming the rights of security, of identity, and of belonging, all of which they have so far been denied. It allows them to understand, on a personal level, the intricacies and implications of their exile, and in recognising that this same sentiment is experienced by fellow Palestinians during their own visits 'home', they are able to establish a sense of unity, of common understanding in exile.

As may be evident, there are echoes of MacCannell's (1999) five stages of site sacralisation present throughout the above. The houses are in many ways named, framed and elevated, enshrined, and reproduced mechanically and socially. Yet ultimately, it is their unique status as both individual and collective sites of exile that marks them as sacralised sites. They have been imbued with a special significance that at once represents the familial connection to the homeland and the *Nakba* itself. Therefore, in visiting their family houses and in recreating parts of them, Palestinians are able to feel that they are connected both to their own family histories and to fellow Palestinians in exile.

Conclusion

As has been noted, MacCannell's (1999) stages of site sacralisation have typically been applied to single locations, such as a castle or a tomb, which are collectively sacralised by large numbers of people who deliberately seek them out. Owing to their beauty, historical, social, or political significance, the sites come to be seen as worthy of special attention, and are marked off from other sites. As their popularity grows, and they incur greater numbers of visitors, the sites, and the structures or places that enshrine them, become essential destinations on travel itineraries, and efforts to reproduce them become widespread. As has been discussed throughout this chapter, however, in instances where it is not possible to visit the site, sacralisation may occur somewhat differently. In the case study of the Palestinian appropriated houses, the majority of the houses have either been appropriated or destroyed, and owing to legal and political impediments may not always be visited. In addition to this, they are scattered across many locations, and each is often only visited by a single individual or family for whom it holds special meaning. Yet, as has been demonstrated, these houses also carry a collective significance for Palestinians.

They are symbols of the *Nakba* and more specifically, of the refusal of the Palestinian people to abandon their right of return. As such, they are perceived not only as the first houses of the first generation of Palestinians in exile, but also as those that should have been their children's and grandchildren's houses. This serves to further strengthen the connection to the appropriated houses, and as has been discussed throughout this chapter, to elevate their status in the minds of both individual families and of Palestinians in exile, collectively. It is precisely this which serves as the basis of the sacralisation of the appropriated houses, and explains why visits to them understandably carry a special significance, both individually and collectively.

The journeys undertaken may therefore be considered as a form of diasporic tourism that occur across many different locations, and yet to a single site. Each house is seen to hold both a particular individual/familial meaning, as well as a collective historical, cultural, and political one. Although both are important, it is the latter that marks them out as a single site. Thus, while each visit to an appropriated house technically occurs in a different place, they nevertheless all occur within the same collective space. As has been noted, these journeys are not undertaken lightly and, in many cases, result in a sense of regret or disappointment. Yet, despite this, for many Palestinians, visiting them remains an important undertaking, one that fulfils a personal and filial duty. In this way, the appropriated houses may be seen as uniting Palestinians living on the inside and on the outside, at once representing the family home, the homeland, and the exile.

References

Abunimah, A., 1998. Crossing the River. *The Link*, 31(5): 1–5.

Al-Rasheed, M., 1991. Invisible and divided communities: Arabs in Britain. In *Selected Papers Presented at the One Day Conference on Arab Communities in Britain, Concerns and Prospects*, held on Saturday 6 October 1990 at University of London, School of Oriental and African Studies. London: Riad el-Rayyes Books, 1–13.

Anderson, B., 1991. *Imagined Communities: Reflections on the Origin and Spread of Nationalism*. London: Verso.

Bachelard, G., 1992. *The Poetics of Space*. Boston: Beacon Press.

Basu, P., 2007. *Highland Homecomings: Genealogy and Heritage Tourism in the Scottish Diaspora*. New York: Routledge.

Bell, C., 2002. The big 'OE': Young New Zealand travellers as secular pilgrims. *Tourist Studies*, 2(2), 143–58.

Blunt, A. and Dowling, R., 2006. *Home*. London: Routledge.

Boullata, K., 2009. *Palestinian Art, 1850–2005*. London: Saqi Books.

Bruner, E.M., 1996. Tourism in Ghana. *American Anthropologist*, 98(2), 290–304.

Bshara, K.A., 2008. *A Key and Beyond: Palestinian Memorabilia in the Economy of Resistance*. Irvine, CA: Center for Global Peace and Conflict Studies, University of California, Working Paper Series.

Bunkše, E., 2004. *Geography and the Art of Life*. Baltimore: John Hopkins University Press.

Burch, S., 2002. Shaping symbolic space: Parliament Square, London as sacred site. In A. Phelps, G.J. Ashworth and B.O.H. Johansson, eds, 2002. *The Construction of Built Heritage: A North European Perspective on Policies, Practices and Outcomes.* Aldershot: Ashgate, 223–36.

Coles, T., and Timothy, D.J., 2004. My field is the world. Conceptualizing diasporas, travel, and tourism. In T. Coles and D.J. Timothy, eds, 2004. *Tourism, Diaspora and Space.* New York: Routledge, 1–30.

Crouch, D., 2003. Writing of Australian dwelling: Animate houses and anxious ground. *Journal of Australian Studies*, 27(80), 43–52.

Darwish, M., 1995. *Memory for Forgetfulness: August, Beirut, 1982.* Berkeley: University of California Press.

Dovey, K., 1985. Homes and homelessness. In I. Altman. and C. Werner, eds, 1985. *Home Environments.* New York: Plenum Press, 33–64.

Farah, R., 1999. Paradoxical and overlapping voices: The Refugee-UNRWA Relationship and Palestinian Identity in Jordan. Paper presented at the International Symposium: *The Palestinian Refugees and UNRWA in Jordan, the West Bank and Gaza, 1949–1999* on 31 August–1 September, 1999. Available at: <http://repository.forcedmigration.org/show_metadata.jsp?pid=fmo:1943> [Accessed 20 January 2014].

Farah, R., 2005. Palestinian refugee children and caregivers in Jordan. In D. Chatty and G.L. Hundt, eds, 2005. *Children of Palestine: Experiencing Forced Migration in the Middle East.* New York: Berghahn, 87–121.

Farah, R., 2007. *Oral history in the Palestinian and Sahrawi Contexts: A Comparative Approach.* Badil Resource Center for Palestinian Residency and Refugee Rights. Available at: <http://www.badil.org/es/educacion-y-activacion-de-jovenes/item/977-oral-history-in-the-palestinian-and-sahrawi-contexts-a-comparative-approach> [Accessed 5 February 2013].

Giuliani, M.V., 1991. Towards an analysis of mental representations of attachment to the home. *Journal of Architectural and Planning Research*, 8(2,: 133–46.

Hammer, J., 2005. *Palestinians Born in Exile: Diaspora and the Search for a Homeland.* Austin: University of Texas Press.

Jackson, M., ed., 1995. *At Home in the World. Sydney: Harper Perennial.*

Jacir, A.K., 2002. Refugees and the right of return. *Socialism and Democracy*, 16(2), 48–53.

Kanafani, G., 2000. *Palestine's Children: Returning to Haifa and Other Stories.* Boulder: Lynne Rienner Publishers.

Kawar, W.K. and Nasir, T., 1980. The Traditional Palestinian Costume. *Journal of Palestine Studies,* 10(1), 118–29.

Khalidi, W., 1992. *All that Remains: The Palestinian Villages Occupied and Depopulated by Israel in 1948.* Washington, DC: Institute for Palestine Studies.

Kraut, A.M., 1982. *The Huddled Masses: The Immigrant in American Society, 1880–1921.* Arlington Heights: Harlan Davidson.

Lennon, J.J. and Foley, M., 2000. *Dark Tourism.* London: Continuum

Löytynoja, T., 2008. The development of specific locations into tourist attractions: Cases from Northern Europe. *Fennia-International Journal of Geography*, 186(1), 15–29.

MacCannell, D., 1999. *The Tourist: A New Theory of the Leisure Class*. Berkeley: University of California Press.

Masalha, N., 2008. Remembering the Palestinian Nakba: Commemoration, oral history and narratives of memory. *Holy Land Studies: A Multidisciplinary Journal*, 7(2), 123–56.

Massey, D., 1994. *Space, Place and Gender*. Minneapolis: University of Minnesota Press.

Matar, D., 2005. *News, Memory, Identity: the Palestinians in Britain*. PhD. London School of Economics, London.

Nagel, C., 2001. Hidden minorities and the politics of 'race': The case of British Arab activists in London. *Journal of Ethnic and Migration Studies*, 27(3), 381–400.

Nassar, H.K., and Rahman, N., eds, 2008. *Mahmoud Darwish, Exile's Poet: Critical Essays*. Northampton: Olive Branch Press.

Nasser, C. D., 2008. *Classic Palestinian Cuisine*. London: SAQI.

Office for National Statistics, 2001. *2001 Census. Country of Birth by Religion*. London: Office for National Statistics.

Porteous, J.D., 1976. Home: The territorial core. *Geographical Review*, 66(4), 383–90.

Ramadan, A., 2009. A refugee landscape: Writing Palestinian nationalisms in Lebanon. *ACME*, 8(1), 69–99.

Rose, G., 2003. Family photographs and domestic spacing: A case study. *Transactions*, 28, 5–18.

Said, E. (with Mohr, J.), 1986 [1993]. *After the Last Sky: Palestinian Lives*. London: Vintage.

Said, E., 2000. *Reflections on Exile and Other Literary and Cultural Essays*. London: Granta Books.

Sayigh, R., 1998. Palestinian camp women as tellers of history. *Journal of Palestine Studies*, 27(2), 42–58.

Shabaneh, G., 2010. Refugees, international organizations and national identity: The case of Palestine. *New Political Science*, 32(2), 215–36.

Shehadeh, R., 2010. *Strangers in the House*. London: Profile Books.

Shakkour, S., 2010. *Christian Palestinians in Britain*. PhD. University of St Andrews.

Skinner, M. and Kawar, W., 2007. *Palestinian Embroidery Motifs: A Treasury of Stitches, 1850–1950*. London: Melisende.

Somerville, P., 1997. The social construction of home. *Journal of Architectural Planning and Research*, 14(3), 226–45.

Tuan, Y.-F., 1971. Geography, phenomenology and the study of human nature, *Canadian Geographer*, 15, 181–92.

Tuan, Y-F., 2004. Sense of place: Its relationship to self and time. In T. Mels, ed., 2004. *Reanimating Places: A Geography of Rhythms*. Great Britain: TJ. International Ltd., 45–56.

Tucker, A., 1994. In search of home. *Journal of Applied Philosophy*, 11(2), 181–7.

Wise, J., 2000. Home: Territory and identity. *Cultural Studies*, 14(2), 295–310.

Yahya, A.H., 1999. *The Palestinian Refugees, 1948–1998: An Oral History*. Palestinian Association for Cultural Exchange (PACE).

Young, I.M., 1997. *Intersecting Voices: Dilemmas of Gender, Political Philosophy and Policy*. Princeton, NJ: Princeton University Press.

SECTION IV

Afterword

Chapter 12
Contemporary Pilgrimage: Journeys in Time and Space

Tim Edensor

Introduction

Collectively, the chapters in this book persuasively testify to the sheer diversity of contemporary mobile practices that are organised around visiting valued sites and moving along cherished pathways. These modern pilgrimages, including those organised around the religious voyages that emerged in earlier times but persist, vary widely in the object of their quest, the forms of mobility undertaken, the distance covered and the time allotted. Travel in pursuit of esteemed venues takes place across varied, multiple landscapes as well as sea and air. Such pilgrimages are part of the broader expansion of myriad forms of travel, cultural practices and tourist destinations that characterise contemporary mobility on a global scale (Hannam *et al.*, 2014). As regions, cities and countries vie to attract visitors for economic gain, proliferating new objects of travel desire emerge to supplement venerable attractions, including former homes or birth places, sites of romantic trysts, sporting venues (Gammon and Fear, 2005), grand festivals, sites of celebrity death (Rojek, 1997) and fan worship (Jasud, 2009), world wonders and a host of others that shape the composition of the modern 'bucket list'. For example, Mecca is developing a huge tourist and shopping infrastructure to serve the increasing numbers of Muslim pilgrims that descend on the city for the annual Hajj, and as well as visiting that hallmark achievement of urban modernism, New York's Empire State Building, tourists are now also drawn to witness the much larger Burj Khalifa in Dubai.

The increasing multiplicity of pilgrimage sites that serve as venues at which to perform diverse practices and express different identities refutes any attempt to draw clear distinctions between tourism and pilgrimage, for as is clear, such journeys are apt to blur and mingle sacred and profane practices. For instance, at the Taj Mahal, crowds of villagers disembark from their coach and collectively make a rather hurried circuit of the monument. Passing through Agra, they are compelled to visit the renowned global icon, and while they rarely perform the typical tourist practice of photography, they take some time to view the mausoleum and grounds. Yet the Taj is not their destination, for they are travelling towards Mathura/Vrindaban, the holy birthplace of Lord Krishna, and here they will participate in more expressive and engaged Hindu religious practices (Edensor, 1998).

This blurring is also generated by the ways in which, as Michael Sallnow (1991) contends, sites of pilgrimage are contested between secular, political, commercial and sacred interests. As Hopkins *et al.* discuss in their account of the Kumbh Mela, different participants make different commitments to the event, with the deep engagements of the Kalpwasis contrasting with the dispositions and practices of day visitors and tourists.

This contestation can be further exemplified at the Taj Mahal, a multiple contested symbolic site (Edensor, 1998). The low intensity of the sojourn around the grounds by the aforementioned parties of villagers strikingly contrasts with the passionate, intense engagements of Western package tourists and many Muslim visitors. Most package tourists have a limited time to spend at the site since they are constrained by the structures of their rigid itinerary. Since for the majority, the Taj is the key attraction in their decision to visit India, guiding their secular pilgrimage to what they commonly regard as a global emblem to love, they inscribe its importance through a highly focused period of concentrated, contemplative gazing and photographing. For Muslim visitors, the Taj possesses a very different significance. Created on behalf of the Moghul emperor Shah Jahan, the site represents a period in which Muslim rule predominated across the sub-continent, but more importantly, the walls are lavishly inscribed with Quranic text, and besides the mosque adjacent to the mausoleum that serves as a venue for prayer, this solicits an ardent religious engagement. Along with the languorous visits of backpackers and the socially convivial parties of middle class Indians, the Taj is the site for different practices, meanings and levels of involvement that are shaped by different understandings of its value as a site for pilgrimage. In developing a conceptual grasp of contestations such as this, along with the sheer diversity of routes, destinations, practices and experiences that characterise contemporary pilgrimage and are exemplified in this book, I now focus on some of the distinctive temporalities and spatialities that shape such processes.

The Temporalities of Pilgrimage

As the editors claim, despite appearances and the intentions of key actors to secure meanings and importance, sites and routes of pilgrimage are rarely fixed and stable but arise from the contingent conjunctions of pilgrims, managers, entrepreneurs and authorities and their practices at particular points in space and time. The varied pilgrimages that form the focus of chapters in this volume highlight the multiple temporalities that shape the sacrality and significance of sites and the experience of travelling to them.

Certain sites may become important and then lose that significance. Christ's Well, located in Blairdrummond in central Scotland, was particularly popular amongst pilgrims who were encouraged to visit the site by the Church. Throughout Scotland, pilgrims would make their way to the well to pray and seek succour from the supposed health-giving properties of its water, leaving behind a piece of cloth or silver as offerings, possibly testifying to the pre-Christian significance of the site. Following the Scottish Church's Reformation in 1581, a campaign against what now became labelled idolatry and superstition was waged by the Kirk, with ministers ordained to compel any parishioners who had made the journey to the well to repent during services. For the next 60 years, penalties for pilgrimage included fines and imprisonment, despite claims about numerous incidences of healing. A century later, there are no records of the well's existence, so successful was the zealous campaign against the practices of pilgrims. It was not until the late eighteenth century that the well's historical significance re-emerged when it was landscaped on the instructions of the local laird, and transformed by the addition of a pool, rockery and standing stone, a romantic, ornamental addition to the estate in an era of grottoes, hermits' caves and fake ruins. Today, these romantic features can still be discerned but the well is gradually melding with the surrounding terrain. Almost unknown, it nestles in the shadow of a wooded hillside adjacent to the nearby safari park (Holder, 2008).

Similarly, routes also alter as Belén Maria Castro Fernández details in her chapter discussing El Camino de Santiago. Fernandez shows how the way has continuously changed through history, with its symbolic importance contested according to political vagaries, and its course altering depending upon the varying influences exerted by commercial interests, businesses, religious fads and secular enthusiasms. Places along the route have continually competed to attract pilgrims and have forged alternative byways and diversions. A key influence on the Camino, Fernandez shows, followed the medieval institutionalisation of the route and its subsequent decline, and General Franco subsequently appropriated the route for his own nefarious political reasons to advance national(ist) values of monumentalism, tradition and order. These symbolic meanings have more recently been challenged by an emphasis on the shared European cultural and historical context of the route, and in contemporary times the journey has become popular amongst secular tourists who seek out an encounter with the past or a testing physical challenge.

A different kind of temporality features in Suha Shakkour's melancholy discussion of those who undertake journeys to Palestine to search for homes that were repossessed by the Israeli state following their enforced exile. This temporal disjuncture, between a sensually charged childhood home and what is the opposite of the notion of a home as refuge, now occupied by Israeli Jews or demolished, serves to rekindle emplaced recollections and stimulate the ongoing emergence of memory and identity. This pilgrimage also testifies to the ways in which the idea of home can be mobile as well as reinforcing or challenging pilgrim's ideals of that first, now appropriated home, which is also a synecdoche of the homeland and fuels the collective narrative of the *Nakba*. The cross-generational power solicited by this process through which the idea of an original home is kept alive foregrounds a claim of right to the home and homeland, nurturing a collective experience of exile and potentially prefiguring a future return. Yet in intensely summoning up the everyday sensations and practices of home that bodies can remember, such visits can foster an enduring pain of separation that cannot be alleviated. As Lucy Lippard considers, '(I)f one has been raised in a place, its textures and sensations, its smells and sounds, are recalled as they felt to child's, adolescent's adult's body' (1997: 34), and in visiting former homes, these pilgrims may be transported back in time. Indeed, many Palestinians choose not to undertake such pilgrimages for transformed settings may violate cherished, sacrosanct memories of former inhabitation.

These Palestinian journeys towards and away from a shared past are different from but resonate with ideas that conceive the pilgrim's journey as a test of belief and a search for faith, in that they produce a passage of time that opens up potential for revelation, inspiration and enlightenment, as well as occasions when one must confront the dark nights of the soul. In her discussion of the extraordinarily dogged voyages on foot undertaken by George Fox, Pamela Richardson reveals how contemporary Quaker pilgrims try to gain a sense of this original, lengthier commitment to spreading faith, attempting to place themselves within his story, to empathetically envision his arduous travels, the religious persecution he endured, the physical endurance required, the social bonds forged and the sensual apprehension of the space traversed. As with the Palestinians who travel back home these pilgrims endeavour to access earlier times; by walking along historical routes, they honour the travails of the founder of their religion and his route to enlightenment or faith while attempting to imaginatively capture that experience.

This engagement with the temporalities of pilgrimage is also importantly undergirded by the focus in some chapters on the various occurrences and stages of the journey itself, the rhythms, flows, stoppages and events that constitute temporal passage. This focuses attention on Tim Cresswell's (2006) assertion that the qualities of the route travelled need to be afforded due recognition so that it is not merely conceived as an abstract temporal phase that occurs between two meaningful points. For the journey commences, unfolds, passes through distinct phases, nears completion and reaches its end. Pilgrimage journeys have a temporal shape within which successive experiences and cultural practices flow that may alternately be purposive, social, self-absorbed or habitual, melding the sacred and profane over time.

As David Crouch insists here, pilgrimage has a becoming quality, and though it may be organised according to all sorts of objectives and timings, numerous contingencies may occur, all sorts of affective entanglements, surprising happenings, emotional intensities and sensual excitations that meld with the flow of time. As Katy Beinart persuasively demonstrates in her chapter, the journey can disclose as much as the destination. It can serve as an opportunity of attunement and reflection about the past and one's ancestors, and encourage a willed focus on giving time to reverie and contemplation. Beinart shows how during the course of her maritime journey in the company of her sister, they engaged in a process of coming to terms with the past, and of making connections with it, but also of considering the future by making contingent, improvisational stories and conjectures that acknowledged their disengagement from the religious dimensions of Jewishness but also foregrounded those elements of Jewish identity that they have retained and performed.

The flow of multiple temporalities and varying dispositions towards time during this journey highlights that though it may often be conceived as an extraordinary voyage, a once in a lifetime venture, pilgrimage also contains much of what is reiterative, ordinary and mundane. As Hopkins *et al.* point out, much of the journeys undertaken by pilgrims are filled with everyday experiences, and with the performance of new kinds of habits that become unreflexive as the excursion progresses. The infrastructural provision of catering and sleeping facilities at destinations and along the way testify to how the very process of travelling necessitates the instantiation of a regular series of practices that may become routine, as do so many of the practices of other forms of travel and tourism (Edensor, 2007). Yet though the highlights of journeys may lie at their culmination, when the sacred place is reached, it is the very condition of routine that opens up the potential for the surprising to emerge, for the everyday is not merely concerned with dull, repetitive action. While habits are essential for social cohesion and a sense of belonging as people together enact familiar manoeuvres (Frykman and Löfgren, 1996), in habitual, everyday performances, as Paul Harrison considers, 'there is always immanent potential for new possibilities of life' (2000: 498) since quotidian practice is open-ended, fluid and generative, replete with daydreams, disruptions and arresting sensations.

Quotidian moments of wonder and fascination that interrupt the flow of the journey are also captured in Katy Beinart's account. During the passage, unpredictable, alienating, mysterious, disturbing or welcoming sensations and memories could be provoked by the books she took to read, or by the very liminality of the ship temporarily moving beyond national borders, conditions that fostered an often playful and sometimes serious, mindful engagement. The emergence of the extraordinary in the routine is particularly vivid when the sisters are initiated into becoming 'rightful persons of the sea', lulled into engaging with the everyday context of their voyage as young women in a typically male, nautical environment.

The Spatialities of Pilgrimage

The examples discussed throughout this collection also foreground the particular geographies into which sites of pilgrimage are enfolded, and draw out the spatial qualities of routes and sites. As with older sacred places, these sites of pilgrimage are constituted through mobile practices that along with the performance of sacred rites at the destination are inseparable from the broader process of attaining wisdom, worldliness, experience or spiritual enlightenment. The links between embarkation points, routes and destinations highlights how all places are continually (re)constituted through their connections with other places (Massey, 1993). In this sense, the differently scaled processes that move between, across, to and from places, continuously remake them, bringing in flows of people, ideas, and finance from elsewhere, and at sites of pilgrimage, such flows may be especially key dynamic processes in their ongoing reconstitution. Since at pilgrimage sites varied routes converge, they often serve as important nodes in larger spatial networks of sacred places or must-see attractions. For instance, the prehistoric places conjoined by ley lines and sites conceived as new age centres of light that further connect to cosmological, divine or ethereal spaces exemplify how sites may be inserted into wider symbolic networks of mobility. So too, the coach tours organised to visit selective sites of national significance reproduce a symbolic constellation of connected places.

A focus on the meaningful, sensual and routine experiences of pilgrims as they move through space towards their destination further demonstrates the inappropriateness of considering journeys of all sorts as spells of empty time through abstract space that simply link two places that are by contrast, replete with meaning. In contradicting such assumptions, several chapters in this collection foreground the ways in which a mobile sense of place is achieved. I have argued elsewhere that even the supposedly alienating, even dystopian, mobile practice of commuting may be replete with familiar rituals and fixtures, unusual sights and happenings, and comfortable sensations (Edensor, 2011) that install a sense of homely belonging to the time and space traversed. The rich experience of travelling during pilgrimage also solicits such experiences.

They are particularly well captured in Nina Vollenbröker's stimulating chapter about the journeys of Americans moving westwards towards the promise of a better life. Vollenbröker shows how these travellers were continuously engaged in place-making activities as they rested and moved through the landscape. Daily rituals of cleaning, cooking and eating had to be performed along the route, as well as the more symbolic practice amongst some women of making quilts. Equally significant was the maintaining of diaries through which they recorded their passage through the landscape, making sense of it through writing and sketching. Travellers situated their journeys in space and time by recording distances and dates, as well as geology, landscape features, natural history and weather, and by naming places, perhaps as a mode of domesticating unfamiliar realms. And while this land could be inhospitable, with travellers suffering from disease and the effects of extreme heat or cold, a spatial immersion was practised that involved the gathering of plants and hunting of animals for food, as a practical knowledge about these places was acquired. Yet these journeys were not merely practically performed for as Vollenbröker shows, in the diaries ecstatic, sensual appreciation of the grandeur of the landscape reverberates through evocative passages that depict its sonic, olfactory and tactile qualities, with descriptions of melodious birdsong, verdant spaces and beautiful vistas, and these emotional engagements further inspired the collecting of mineral and organic souvenirs. All these activities, she argues, nourished the emergence of a rich, thick sense of place grounded in a mobile, ongoing entanglement with the land.

These rich experiences of the journey conjure up speculations about the centrality of rootedness to Western thought and the dominant values that surround the sedentary inhabitation of place. On the travels discussed by Vollenbröker, a contrasting sense of belonging is found in mobility, perhaps resonating with Bruce Chatwin's (1997) thesis that the pilgrimage is a social strategy institutionalised to temporarily escape the constraints of sedentarism. According to Chatwin, the notion of rootedness in one place is inimical to human desires and dispositions. The sedentary life is of only recent historical provenance since for most of human history mobility has been prevalent, best placed to stimulate the brain, contribute to a sense of well-being and develop understanding of others.

In foregrounding the dynamic, becoming qualities nurtured by mobility however, there is a danger that place may by contrast be misconceived as a reified, static realm, ossified and unchanging, in contradistinction to Massey's (1993) insistence that places are forever being recomposed out of their links with elsewhere. Vollenbröker's discussion of travellers' immersive engagements with animated landscapes and places points to the vitality of places and landscapes, and of the natural and social practices which take place in and around them, thus refuting notions of fixity. This is further exemplified in several of the chapters in which the spatial affordances confronted by pilgrims require embodied, mindful engagements that invoke notions of the magical or sacred.

These animated qualities of the landscape and their interlacing with the symbolic and experiential dimensions of the sacred is eloquently captured in James Thurgill's essay on the Pagan New Year festival of Samhain at the venerable Ankerwycke Yew. The yew's charismatic dimensions impact upon the embodied senses of pilgrims, and its extraordinary size combined with the twisted signs of its ageing and the material adornments on its surface draws them in to perform ritual. The sacrality of the site and the centrality of the tree to it is intensified by the affordances of its situation, since water filled ditches and adjacent lines of trees form a border to enclose the quiet, secluded place in which the ritual occurs. For pilgrims, this sensual, sealed space fosters an apprehension of other, invisible preternatural forces that course through the world, animated flows that lie beyond those usually sensed. This may be conceived as the agency of the tree in drawing up water and sustenance into its body, but might also include more evanescent natural energies and mystical flows that combine to consolidate the hallowed significance of the site as spiritually enchanted.

Maddrell and Della Dora (2013) point out how pilgrims' interpretations and experiences of the landscape through which they travel refer to divine resonances that differ from the aesthetic, romantic gaze mobilised by other onlookers. Pamela Richardson brings out the ways in which the affordances of the landscape are endowed with such spiritual associations. On the Quaker paths of George Fox, a lofty hill affords a prospect across a landscape that can be conceived as evocative of a divine sublimity, and a sheltered valley can be construed as affording safety and restoration to the travelling body. The prevalence of holy wells throughout pre-Christian and Christian Europe testifies to the healing powers invested in the pellucid water that gushes from underground springs and the mooted sacred qualities of the river Ganges during the Kumbh Mela cleanses and purifies the pilgrims as they ritually bathe.

These symbolic and animated landscapes underscore the potential for wonder to emerge during pilgrimage. As Jane Bennett remarks, to become enchanted is 'to be struck and shaken by the extraordinary that lives amid the familiar and everyday', for instance through 'a momentarily immobilising encounter; it is to be transfixed, spellbound'. (2001: 4). The sacrality of a site is thus usually imbricated with its more-than-representational qualities, with a range of sensations that enliven the body and with the flow of atmosphere (Edensor, 2012). Pilgrims are likely to enter a sacred place already attuned to receive such affects, opening their bodies out to sensory, spiritual ritual experience, ready to be enchanted. The pilgrim's disposition towards enchantment and to becoming entangled with particular kinds of animated landscapes and places interrogates scientific approaches to establishing truth and veracity. Such rational missions operate in an entirely different register and misrecognise the realm of the sacred. In his chapter, Rob Irving recognises this in his construal of the 'mythic', a realm of understanding that is imaginary yet provides a basis for speculative knowledge. The engagement with sacred space, he contends, is a creative, collaborative process that produces a form of understanding which sidesteps rational and scientific enquiry.

This collaborative endeavour by pilgrims to cultivate an experience of the magical or sacred, or to achieve a sense of *communitas* with each other, highlights the highly social nature of pilgrimage, for though there are those who set off on a solitary journey of self-discovery, most pilgrims are part of groups that experience events during their passage collectively. In theories of tourism, far too many accounts construe a single, purposive subject who is involved on some individual quest for meaning and experience and yet most tourist endeavours are replete with social interaction, involve taking care of family members, of consolidating friendship bonds and of sharing key experiences along the way (Larsen, 2008; Blichfeldt and Mikkelsen, 2013). Similarly, in the context of pilgrimage such social engagements are widely discussed in the chapters in this book.

As well as being a co-production with others, pilgrimage usually depends on the presence of others, a form of being *with*. This is particularly evident at a huge event such as the Kumbh Mela, where millions descend on the campsite, pray, eat and bathe together. In their chapter, Maddrell, Terry, Gale and Arlidge demonstrate the intense sociality of those drawn to attend the annual TT motorcycle races in the Isle of Man, an event where the co-presence of others is integral. Many attendees repeatedly make the trip each year, with old friends, and each year carry out regular practices pertinent to the occasion, performing collectively to re-establish the sense of *communitas*, and confirm shared meanings and values.

These collective social performances highlight how sacred places and routes are reconstituted through the action of pilgrims, securing their significance and symbolic values. RPB Singh (1987) shows how the Hindu pilgrims that descend on the sacred Indian city of Varanasi follow the proliferating sacred routes that cross the city, performing symbolic rituals that integrate cosmological meanings, mythic value and social and embodied experience. Moreover, the sacred symbol of *mandala* is superimposed upon the material form of Varanasi and, enacted by pilgrims through these ritual passages, reinscribes religious identity in and on space. Varanasi also includes within its sacred geography physical sites that symbolise other sacred places within India and the ritual performances of pilgrims sustain resonances that expand across national and divine spaces.

Yet though ritual and its reiteration are essential for the consolidation of meaning over time, as David Crouch points out, this must not be considered to prevent an openness to possible consequences and the ability to improvise. Katy Beinart and her sister engaged in improvisatory practices during their voyage, eating the distinctive foods and wearing the clothes of their forebears so as to try and empathetically sense their embodied experiences of travel. Though conjectural, such manoeuvres opened up the potential for experiencing mobility and the world otherwise. As I have insisted, places and routes are contested and meanings and practices associated with them change over time, adapting and changing their significance. Pilgrimage is therefore a hugely dynamic process that sustains the significance of certain sites, forgets the sacrality of others, brings new places into the orbit of the sacred and transforms and contests the meanings of other sites and routes. In the present, this dynamic scenario operates at an increasingly global level and expands to incorporate an ever increasing range of places.

The expanding diversity of sites and routes, the varied motivations and goals of pilgrims, the contestations over practice and meaning, and the ever changing feelings and meanings that infuse journeys and significant sites thwart attempts to forge over-general theories. The multiple experiences of pilgrims that range from the mundane to the extraordinary include social, spiritual and sensual engagements with people and places, and a host of contingencies and improvisations similarly deter simplistic encapsulations. Until recently, tourist studies has been subject to the influence of hugely over-general theories that suggest that tourism has a particular function, for instance, by serving desires for a temporary quest for liminality (Turner and Turner, 1973) or the experience of authenticity (MacCannell, 1976), or is typified by a particular practice such as gazing (Urry, 1990). Such notions seem wholly misplaced in the face of the enormous diversity of destinations, practices and identities that continually reconfigure tourist experience. In highlighting a similar multiplicity, the chapters in this book argue for a more nuanced, contextual understanding of pilgrimage.

References

Bennett, J., 2001. *The Enchantment of Modern Life: Attachments, Crossings, and Ethics*. Princeton: Princeton University Press.

Blichfeldt, B. And Mikkelsen, M., 2013. Vacability and sociability as touristic attraction, *Tourist Studies*, 13(3), 235–50.

Chatwin, B., 1997. *Anatomy of Restlessness*, London: Picador.

Cresswell, T., 2006. *On the Move: Mobility in the Modern Western World*, London: Routledge.

Edensor, T., 1998. *Tourists at the Taj*, London: Routledge.

Edensor, T., 2007. Mundane mobilities, performances and spaces of tourism, *Social and Cultural Geography*, 8(2): 199–215.

Edensor, T., 2011 The rhythms of commuting, in T. Cresswell and P. Merriman (eds) *Mobilities: Practices, Spaces, Subjects,* Aldershot: Ashgate

Edensor, T., 2012. Illuminated atmospheres: Anticipating and reproducing the flow ofaffective experience in Blackpool, *Environment and Planning D: Society and Space*, 30, 1103–22.

Frykman, J. and Löfgren, O.,eds, 1996. Introduction, in *Forces of Habit: Exploring Everyday Culture*, Lund: Lund University Press.

Gammon, S. and Fear, V., 2005. Stadia tours and the power of backstage, *Journal of Sport and Tourism*, 10(4), 243–52.

Hannam, K., Butler, G. And Paris, C., 2014. Developments and key issues in tourism mobilities, *Annals of Tourism Research*, 44, 171–85.

Harrison, P., 2000. Making sense: Embodiment and the sensibilities of the everyday, *Environment and Planning D: Society and Space*, 18, 497–517.

Holder, G., 2008. *The Guide to Mysterious Stirlingshire*, Stroud: The History Press.

Jasud, L., 2009. St Elvis, *Critical Inquiry*, 35(3), 440–48.

Larsen, J., 2008. De-exoticising tourist travel: Everyday life and sociality on the move, *Leisure Studies,* 27(1), 21–34.

Lippard, L., 1997. *The Lure of the Local: Senses of Place in a Multicentered Society*, NewYork: The New Press.

MacCannell, D., 1976. *The Tourist*, London: Macmillan.

Maddrell A. and Della Dora, V., 2013. Crossing surfaces in search of the Holy: Landscape and liminality in contemporary Christian pilgrimage, *Environment and Planning A*, 45: 1105–26.

Massey, D., 1993. Power geometry and a progressive sense of place, in J. Bird, B. Curtis, T. Puttnam and L.Tickner eds, *Mapping the Futures: Local Cultures, Global Change*. London: Routledge.

Rojek, C., 1997. Indexing, dragging and the social construction of tourist sites, in C. Rojek and J. Urry eds, *Touring Cultures: Transformations of Travel and Theory*. London: Routledge.

Sallnow, M., 1991. Introduction, in J. Eade and M. Sallnow, *Contesting the Sacred: The Anthropology of Christian Pilgrimage*. London: Routledge.

Singh, R., 1990. The pilgrimage manadala of Varanasi (Kashi) A study in sacred geography, in L.Gopal andD.Dubey eds, *Pilgrimage Studies: Text and Context*. Allahabad: Society of Pilgrimage Studies.

Turner, V. And Turner, E., 1973.*Image and Pilgrimage in Christian Culture*. New York: Columbia University Press.

Urry, J. (1990) *The Tourist Gaze*. London: Sage.

Index

References to figures are shown in *italics*. References to footnotes consist of the page number followed by the letter 'n' followed by the number of the footnote.